Mobile Apps
for Museums

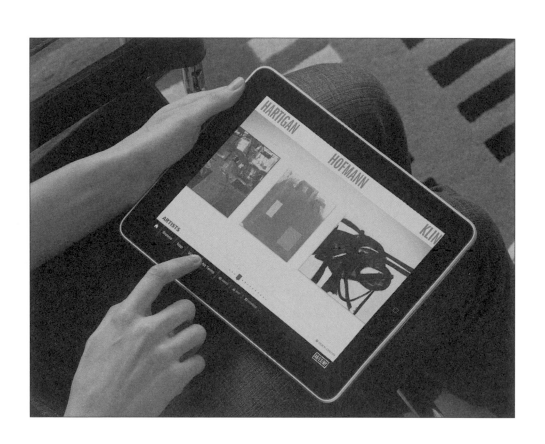

# Mobile Apps for Museums

## The AAM Guide to Planning and Strategy

Edited by Nancy Proctor

With contributions from
Jane Burton, Allegra Burnette, Ted Forbes,
Kate Haley Goldman, Ann Isaacson,
Sheila McGuire, Ed Rodley, Peter Samis,
Scott Sayre, Margriet Schavemaker, Koven Smith,
Robert Stein and Kris Wetterlund

The AAM Press
American Association of Museums
Washington, DC
2011

Library of Congress Cataloging-in-Publication Data

Mobile apps for museums : the AAM guide to planning and strategy / edited by
Nancy Proctor.

  p. cm.

Includes index.

ISBN 978-1-933253-60-2 (alk. paper)

1. Museums—Technological innovations. 2. Museum visitors—Services
for—Technological innovations. 3. Museums—Educational aspects. 4. Mobile
computing. I. Proctor, Nancy, 1968-

AM7.M63 2011

069'.10285--dc23

                    2011017443

Cover and frontispiece photograph courtesy of the Museum of Modern Art, New York

Design: Polly Franchini

# Contents

| 1 | NANCY PROCTOR |
|---|---|
| **Introduction: What is Mobile?** | |

TODAY apps and smartphones probably come to mind first as the iconic, ground-breaking mobile platforms poised to transform the museum experience for all of us. But in fact mobile technologies have been part of the museum landscape since at least 1952 when what may have been the first audio tour was introduced at the Stedelijk Museum in Amsterdam using radio broadcast technology.[1]

Audio tours are still the most common form of "self-guided" mobile experience at cultural sites. Arguably, they are also the oldest source of "augmented reality"(AR), enabling us to "overlay" the observed environment with interpretation and other content we hear. In this light there is a pleasurable echo to finding the Stedelijk once again leading the field in AR apps, discussed here in a chapter by Margriet Schavemaker, the museum's head of collections and research. The Stedelijk example and museums' long history of working with mobile technologies suggests that the foundational experiences and expertise required to deploy even the most cutting-edge of 21st-century mobile technologies effectively lie at museums' fingertips and well within their traditional purview. This introductory volume aims to help museums grasp some of the mobile skills and opportunities most immediately available to them.

Since the invention of the audio tour, the number and kind of mobile devices used by museums have proliferated. Other than audio tours loaned out on made-for-museum devices, podcasts are probably the most common mobile media being published by museums, alongside other kinds of downloadable content ranging from PDFs to eBooks and videos. In terms of personal mobile devices, the majority of the museum's actual and potential audiences still use "dumbphones" that are limited to voice and text messaging. Hundreds if not thousands of museums have created audio tours for this low-cost platform in the past six years or so. What these forms of mobile media—the traditional audio tour, the cellphone tour, the podcast and similar downloadable content—have in common is that they are typically deployed in a broadcast delivery mode: primarily for one-way delivery of content from museum to consumer.

But with today's new networked mobile devices—smartphones, tablet computers and Wi-Fi-enabled media players—two-way communication models are now easier and on the rise. Not just "narrowcast" audio tours but interactive mobile multimedia, including games, crowdsourcing activities, and social media, can be delivered via apps to the visitor's own Internet-enabled phones and media players, instead of or to supplement devices provided on-site by the museum. The term "mobile" has come to encompass an ever-expanding field of platforms, players, and modes of audience engagement. Mobile today means:

- **Pocketable** (phones, personal media players, gaming devices) and **portable** devices (tablets and eReaders);

- **Smartphones** that run apps and access the Internet, and older **cellular phones** that do nothing more than make voice calls and send text messages;

- **Podcasts** of audio and video content, and other **downloadable content,** including PDFs and eBooks;

- **Mobile websites,** optimized for the small screen and audiences on the go, and **"desktop" websites,** designed for large, fixed screens but which are increasingly visited by mobile devices;[2]

- **BYOD** (bring your own device) mobile experiences, designed for visitors' personal devices, and traditional **on-site device distribution** for visitors who do not have or do not care to use their own phone or media player.

Mobile's disruptive power comes from its unique ability to offer the individual intimate, immediate and ubiquitous access combined with an unprecedented power to connect people with communities and conversations in global, social networks: mobile is both private and public, personal and political. Understanding that the new mobile devices today are also geo-spatially aware computers capable of supporting research, communication and collaboration challenges us to "think beyond the audio tour" and our silo-like approaches to digital initiatives. It also inspires us to reinvent the museum's relationship with its many publics by conceiving content and experiences that operate across platforms and disciplines, both inside the museum and beyond.

At the same time that the rise of mobile reshapes the museum's thinking about its digital interfaces, it broadens access to the museum exponentially. Not only are more people able to connect with the museum through their mobile devices, but there is also the potential for them to personalize their museum experience whenever and wherever they like, integrating collections, exhibitions and other offerings into a much broader range of use-case scenarios than we have ever imagined. The museum can not only enter people's homes and classrooms, but can also be part of their daily commutes, their international travel, their work and leisure activities as never before. How will museums understand and cater to this huge range of contexts and demands for cultural content?

---

## Mobile is Social Media

As Koven Smith has argued[3], delivering what is fundamentally the same, narrow-cast audio tour experience to shiny new gadgets is unlikely to improve the take-up or penetration rates of mobile technology used by museum visitors: in other words, to better help the museum deliver on its educational and interpretive mission. Although in conflict with visitors' self-reported usage of mobile interpretation in museums[4], the traditional audio tour reaches a sobering minority of the museum's on-site audiences, whether the tour is provided on made-for-museum audio devices on-site, or accessed through visitors' personal phones or media players. In the pages that follow, Kate Haley Goldman helps us understand this phenomenon in the context of recent major studies of mobile adoption by museums and their visitors, and frames important new questions for future research to guide ongoing developments in the field.

Thinking beyond the audio tour model, Ed Rodley provides tips on how to integrate mobile into the overall museum experience design to create more authentic, compelling and higher quality mobile programs. Jane Burton tackles the new field of "serious mobile gaming" for museums, and Margriet Schavemaker demonstrates how augmented reality can explode the museum experience into new dimensions and territories for artists, curators and exhibition designers, as well as for museum audiences. No less revolutionary is the impact of new platforms on the centuries-old docent or museum guide format: Scott Sayre, Kris Wetterlund, Sheila McGuire and Ann Isaacson describe how iPads and similar tablet computers can transform the live-guided group tour into a multi-platform, multimedia experience.

Museums are also asking how well content designed with the on-site visit in mind can fulfill the needs of those audiences who will never be able to come to the museum in person. Allegra Burnette provides an introduction to cross-platform thinking that optimizes museums' mobile apps for both the on-site visit and beyond. Similarly, Koven Smith's essay on the "roll-out" of mobile programs shows how new marketing approaches can be integrated into mobile project design to reach target audiences more effectively—even if the app is not built or even commissioned by the museum.

Concerns about the impact of mobile programs have always been intertwined with financial and budgetary considerations for museums. Speaking from more than a decade of experience working both in-house and with mobile vendors, Peter Samis lays out all the elements of mobile content production and their business model considerations to help museums make the best choices in the expanding field of mobile products and services. Ted Forbes guides museums through the decision-process of "native vs. web app," and Rob Stein offers a solution for "future-proofing" mobile tours to make them more compatible across the proliferating platforms and devices now available. My own essay on mobile business models examines the new revenue streams that have entered the museum field with new mobile platforms and players in the market, and suggests metrics appropriate to measuring the success of museums' mobile businesses.

Whether audio tour, "un-tour,"[5] "de-tour," or "para-tour," the approaches to museum apps described in this volume aim to go beyond the "narrow-cast" visitor services model. These essays position mobile as an integral part of a web of platforms that connect communities of interest and facilitate conversations among our audiences as well as with the museum itself: mobile is social media. As an indispensible part of the 2.0 museum, mobile supports the key indices of the museum's success vis-à-vis its core mission and responsibility to the public good:

- **Relevance:** the museum's responsibility to make its collections, content and activities meaningful and accessible to the broadest possible audiences;

- **Quality:** the museum's mission to collect, preserve and interpret the invaluable artifacts and key stories, ideas and concepts that represent human culture and creativity;

- **Sustainability:** the museum's enduring obligation to deliver both quality and relevance to its audiences—forever.

The quality and relevance of the museum's discourse are the preconditions for its sustainability, and enable "network effects" that grow audiences and foster self-perpetuating conversations about the museum's collections, activities and messages. Mobile products and services do not yield these benefits on their own, but rather as an integral part of the eco-system of platforms that now make up the museum as "distributed network."[6]

We hope these essays help strengthen the museum network and cultivate stronger connections among our colleagues as we collectively map the important new terrain of mobile in museums. Recognizing that the only constant in the mobile field is change, this publication is designed with expandability and updates in mind: the digital versions include interactive elements that the entire museum community can contribute to, including product design principles and FAQs. New essays will be added to reflect the changing body of knowledge in the mobile field, beginning with chapters on best practice in content development and collaborative production strategies from Sandy Goldberg and Alyson Webb, among others still being planned. Our strategy is to cast the net widely, tapping both veterans and new thinkers in the field, and to mine the museum community's collective experience deeply, in order to yield the guidelines and examples that will enable us all to integrate mobile products and services most effectively and efficiently into the museum of the 21st century.

## Notes

1. Loïc Tallon http://musematic.net/2009/05/19/ about-that-1952-sedelijk-museum-audio-guide-and-a-certain-willem-sandburg/

2. A recent Pew Internet survey indicates that 40% of American adults already had access to the Internet from a mobile phone in 2010 (Smith, 2010). Gartner predicts that by 2013 mobile phones will overtake desktop computers as the most common method for accessing the Internet worldwide. (Gartner, 2010). A 2011 infographic from IBM suggests that the majority of Internet use will be from mobile devices by 2014. Sarah Kessler, IBM Infographic "Mobile by the Numbers" 23 March 2011 http://mashable.com/2011/03/23/mobile-by-the-numbers-infogrpahic/?utm_source=feedburner&utm_medium=feed&utm_campaign=Feed:+Mashable+%28Mashable%29 Accessed 3 April 2011.

3. Smith, K., "The Future of Mobile Interpretation." In J. Trant and D. Bearman (eds). *Museums and the Web 2009: Proceedings*. Toronto: Archives & Museum Informatics. Published March 31, 2009. Consulted October 25, 2010. http://www.archimuse.com/mw2009/papers/smith/smith.html

4. Petrie, M. and L. Tallon, "The Iphone Effect? Comparing Visitors' and Museum Professionals' Evolving Expectations of Mobile Interpretation Tools." In J. Trant and D. Bearman (eds). *Museums and the Web 2010: Proceedings*. Toronto: Archives & Museum Informatics. Published March 31, 2010. Consulted October 25, 2010. http://www.archimuse.com/mw2010/papers/petrie/petrie.html

5. Notes from the "Un-tour Unconference" session, Museums and the Web 2010. Consulted 15 October 2010. http://conference.archimuse.com/forum/untour_unconference_session

6. Proctor, N. "Mobile Social Media in the Museum as Distributed Network," forthcoming in Interactive Museums, ed. MuseumID, London, 2011.

## References

Gartner. (2010) "Gartner Highlights Key Predictions for IT Organizations and Users in 2010 and Beyond." January 13, 2010. Consulted January 27, 2011. http://www.gartner.com/it/page.jsp?id=1278413

Petrie, M. and L. Tallon, "The Iphone Effect? Comparing Visitors' and Museum Professionals' Evolving Expectations of Mobile Interpretation Tools." In J. Trant and D. Bearman (eds). *Museums and the Web 2010: Proceedings.* Toronto: Archives & Museum Informatics. Published March 31, 2010. Consulted October 25, 2010. http://www.archimuse.com/mw2010/papers/petrie/petrie.html

Proctor, N. "Mobile Social Media in the Museum as Distributed Network," forthcoming in Interactive Museums, ed. MuseumID, London, 2011.

Proctor, N. et al. Notes from the "Un-tour Unconference" session, Museums and the Web 2010. Consulted 15 October 2010. http://conference.archimuse.com/forum/untour_unconference_session

Smith, A. (2010) "Pew Internet & American Life: Mobile Access 2010." July 7, 2010. Consulted January 27, 2011. http://www.pewinternet.org/Reports/2010/Mobile-Access-2010.aspx

Smith, K., "The Future of Mobile Interpretation." In J. Trant and D. Bearman (eds). *Museums and the Web 2009: Proceedings.* Toronto: Archives & Museum Informatics. Published March 31, 2009. Consulted October 25, 2010. http://www.archimuse.com/mw2009/papers/smith/smith.html

Tallon, Loïc. "About that 1952 Sedelijk Museum audio guide, and a certain Willem Sandburg," Musematic, May 19, 2009. http://musematic.net/2009/05/19/about-that-1952-sedelijk-museum-audio-guide-and-a-certain-willem-sandburg/. Consulted January 30, 2011.

| 2 | NANCY PROCTOR |
| **Mobile Business** | |
| **Models in a** | |
| **2.0 Economy**[1] | |

MOBILE is changing the way museums do business—whether they are aware of it or not. As "the people formerly known as the audience"[2] increasingly expect information and experiences on demand, whenever and wherever they are, the market is growing for mobile products and services for and about museums. At the same time, museums are beginning to think of audiences in a more granular way, recognizing more variation in needs and interests among their visitors, partners and collaborators, both online and on-site. As a result, museums aspire to create mobile products and service that better suit specific audience groups and contexts. In response, a wide range of new players has entered the mobile scene, loosening the grip of the handful of audio tour companies who dominated the field for over 50 years. Start-up app companies and smartphone manufacturers, mobile network providers and social media gurus, students, freelance content developers, open data protagonists, "citizen curators," new alliances among cultural organizations to co-create content—it seems that nearly every day new forces are emerging to radically reconfigure both the museum mobile landscape and its business models.[3]

A review of the business strategies developed over the 60-year history of the audio tour[4] shows that some persist in this Mobile 2.0 economy; others are the result of new functions possible on handheld computing platforms and the new business interests that are bringing them to market. What hasn't changed is that

·

whether "free" or "paid for" by the end-user, there is always a cost to the museum to develop a mobile program. Following is a brief discussion of some of the main models museums are adopting to pay for their new mobile products and services, while also achieving their educational, outreach, and interpretive goals.

### Omnibus

One way museums have struck a balance between their missions and the need for revenue has been to tie less popular mobile programs—usually the permanent collection audio tours—to the more profitable blockbuster exhibition audio or multimedia tours in "omnibus" contracts with (generally larger) tour companies. In this kind of deal, mobile interpretation of the permanent collection is effectively subsidized by the higher revenues from temporary exhibition tours, which "sell" better. The blockbuster tours are rented to visitors at a relatively high take-up rate (usually over 15% of visitors take the tour, up to 85% or more in the most successful tours), and some percentage of the profit from the temporary exhibition tour sales is plowed back in to creating permanent exhibition tours: in effect, the audio tour vendor is incentivized to produce the less profitable permanent collection audio tour in exchange for the opportunity to manage the more lucrative and PR-worthy blockbuster exhibition contracts. Some economies of scale can be achieved by using the same hardware and distribution infrastructure for both, but profit margins are reduced by piggybacking the permanent collection program on the more mass-market blockbuster tours. For the app incarnation of this business model, see "Freemium."

### Freemium

The new "freemium" model of the Web 2.0 economy has been greeted with enthusiasm by many cultural professionals (see, for example, DaPonte 2010), but so far there are very few examples in the museum market. The National Constitution Center's app has basic visiting information, the Constitution, and links to current political news; it is free to download, but then charges $0.99

for each themed tour within the app, offering free sample stops for each one. MoMA has just introduced its "MoMA Books" app for the iPad, which is free to download but then requires an in-app purchase to download a complete book from the MoMA library. A variant on the omnibus model, freemium combines the concepts of the "free" and "premium" content in one digital product: educational, outreach and revenue imperatives are balanced by providing the app with some amount of free "loss-leader" content (e.g., from the permanent collection, or, as in the case of the MoMA Books app, sample chapter content), with the possibility to make an "in-app purchase" to add more content at a fee (e.g., for the special exhibition; the full book).

### Subscription

In February 2011, iTunes introduced a subscription model for its app-based content. Launched with The Daily news iPad app from News Corporation, this model enables publishers to sell an app that will be updated periodically with new content at a recurring fee to the end-user. This suggests interesting new possibilities for museums to offer not just digital magazines through iTunes, but also subscriptions to apps and digital catalogues or other "ePubs" that are regularly updated. There are two catches: Apple takes 30% of subscription revenue (the same amount they garner from every app sale through iTunes), and perhaps more significantly for museums, these new products may require a new workflow to support the process of updating content. In the typical print production model, once a product has been published, only new editions and print runs permit content changes. Will museum book and catalogue teams be willing and able to adopt a magazine-style process in order to attract additional revenues beyond the initial product sale? Only time will tell, but it is certain that periodical publishers everywhere will be looking for alternatives to iTunes for distributing their digital products so they can cash in on the recurring revenues of the subscription model without having to pay such a hefty commission to Apple. As a result, museums can expect the number of online distribution channels for their downloadable digital products to increase in future.

*Open Data*

In response to an earlier essay on which this one is based and which was published as part of the proceedings for Museums and the Web 2011, Glen Barnes from app company MyTours suggested an additional new business model: making the museum's data and content available to third parties to develop mobile apps and other products "for" the museum and its audiences.[5] While museums might justifiably keep their physical collections under lock and key for their own and the longer-term public good, they are being asked with increasing frequency to open their data and digital collections for use by others. The White House's Open.Gov initiative calls for greater openness and transparency by the Federal Government in the United States, and includes a directive to federal agencies, which includes federally-funded museums, to publish data online,[6] as well as a strategy for making data more accessible, and more data available.[7]

Barnes argues that the benefits outweigh the loss of control for museums, and offers these examples of possible outcomes of museums opening their data:

• A company with an existing tour app could aggregate tours from various museums into one app that allows users to find all of the nearby museum tours. This could open the possibility of your content being found by people who might not otherwise know about your museum.

• Nokia fans/hackers are annoyed that none of the apps for museums are coming out for the Symbian platform...They develop an app that makes the content available on their device.

• Of course it can't all be roses... A content farm scrapes the content, republishes it and wraps a bunch of [ads] around the content.

In his essay in this volume, Koven Smith argues that museums have more control than they may think in the "Open Data" model: by controlling what data they release and in what format—and through a judicious use of creative commons copyrights—museums can influence to a large extent the kind of mobile products that are created with their content.[8] The Brooklyn Museum conducted an early experiment with this model, enabling Iconoclash Media to develop an app for the collection with the Museum's Open Collection API.[9] By inspiring new kinds of partnership and revenue streams, this model offers perhaps the most potential for business innovation in the mobile sector.

### Sponsorship

Another traditional model is sponsorship, which covers part or all of the cost of the mobile program's creation and distribution, allowing the museum to offer the product or service to visitors without charge or at a reduced fee. This model may also free up the mobile content for broader distribution without being in conflict with a revenue source for the museum. In the case of museum tours, the usage rate for the sponsored tour is significantly higher than those without subsidy. For example, when MoMA's audio program was offered for free after the museum reopened in 2004, courtesy of a grant from Bloomberg, the tour usage rate went from 5–8% of visitors to 31%, and now reaches more than 45% with the broader distribution through MoMA Wi Fi and other channels. Similarly, for several years, AT&T sponsored SFMOMA Artcasts, the museum's podcast series (though that sponsorship was eventually transferred to a higher ticket funding opportunity). (Burnette, 2011)

### Advertising-Supported

Until now, sponsorship has been the limit of the introduction of commercial brands into museum settings, unlike symphonies, theater, and ballet, which distribute programs where advertising is plentiful. One example of an early experiment with in-app advertising is the Indianapolis Museum of Art's 100 Acres tour. The museum agreed to a "media-swap" with the local Sunday paper: in exchange for publicity for the mobile website in the newspaper, a "disappearing" banner ad is seen briefly at the bottom of the map each time the map is accessed in the app.

Outside the museum world, some app and mobile website publishing platforms that are free to the publishers and/or users include the platform owner's logo, advertising, or the possibility of selling advertising as an indirect way of paying the platform authors/owners for the use of their service (e.g., WireNode, Mobify; many Twitter apps are free to end-users but include advertising banners). But so far this funding source has not been fully explored by the museum community.

### Membership Benefit

Some museums use audio tours, provided at a fee to most visitors, as a membership benefit. The Royal Academy in London, for example, provides free audio tours along with free entry to their exhibitions for members, enabling them to skip the line to get into blockbuster shows, as well. This is a good example of leveraging a mobile product for its "network effects": in addition to its direct revenue potential, the audio tour adds an incentive to join the museum and thereby drives an additional revenue stream above and beyond tour rental fees.

### Donations

Mobile giving got its biggest boost in public awareness from the Red Cross's text donations to Haiti campaign: in less than 10 days, over $30 million had been donated in $10 increments from 3 million unique donors, with additional donor development benefits for the non-profit: 95% of the text-message donors were "first-time donors to the American Red Cross," and "20,000 opted in to receive ongoing email communications from the nonprofit organization." (Mobilemarketer, 2010) These spectacular results attracted many museums to try mobile giving, but, as Megan Weintraub, new media manager for Oxfam, said to the NonProfit Times, "Not everybody is the Red Cross. You don't have Michelle Obama telling you to text with other organizations." The decidedly more modest results from mobile giving at cultural organizations have yielded the most with event-driven campaigns that make extensive use of traditional marketing and advertising outlets to publicize the cause and its donation short code. The Philips Collection solicited text message donations of $5 and $10 to help restore the museum after the September 2010 fire. (ArtInfo, 2010) Anyone contemplating mobile giving campaigns must also take into account the marketing overheads required to make them successful: signage, traditional media publicity for the campaign, and staff time are all costs to be weighed against new donation revenues. Nonetheless, mobile giving offers benefits beyond just money: increases in new donors, members and opt-ins to the museum's mailing list should also be factored in as potential additional "network effects" that can result from a well-designed campaign.

### The Value Is in the Network

In the summer of 2010, Fast Company blogger Aaron Shapiro wrote:

> Apple CEO Steve Jobs has said, the App Store has generated more than $1 billion in revenue for developers. That sounds like a big number. But ... [o]ne billion dollars in revenue for the approximately 225,000 apps is $4,444 per app—significantly less than an app costs to develop... A typical iPhone app costs $35,000 to develop. The median paid app earns $682 per year after Apple takes its cut. With these calculations for the typical paid app, it takes 51 years to break even. It's not any better for free apps. A free app also costs about $35,000 to develop. But there are so many free iPhone apps that at a rate of 2 seconds per app, it would take approximately 34 hours for someone to check out each one. That's not great odds for a revenue model based on advertising. (Fast Company, 2010)

It is becoming clear that museums are as unlikely as any other developer to "get rich quick" on mobile apps. But scant financial returns on mobile products are really nothing new to the museum field: with a few exceptions among the most visited cultural attractions, revenues have not been the most significant benefits to the museum from its audio tours and their progeny. Nonetheless, museums have been early adopters and innovators on the mobile space for some 60 years. The investment required for mobile programs has commonly been justified because of mobile's unique ability to meet other needs of the museum's mission: offering greater possibilities for extending outreach, improving the quality and accessibility of interpretation and education, and supporting other revenue initiatives and connecting platforms to create a whole greater than the sum of its parts.

The metrics of success for mobile, like its goals, are therefore not just the number of downloads and dollars received, but also the extent to which the mobile program is able to engage audiences and support other museum programs, activities and revenue streams. These outcomes are clearly much more difficult to quantify, but devising metrics, measuring tools, and a management culture that evaluates and values them should be a focus of effort by the museum community as we experiment with new mobile business models. As Max Anderson has indicated, the "network effects" possible when mission-driven initiatives are connected in a healthy eco-system show that there is more than just "red ink" to the business of mobiles in museums. (Anderson, 2007)

# Mobile Product Development Principles

Nancy Proctor
proctorn@si.edu

## Product Development Principles

1. Mobile projects should expand and create new opportunities for engagement, not seek to reproduce existing ones on mobile devices.

2. Mobile should be understood as social media. Projects should leverage social media's ability to create conversations, communities, and collaborations, both alone and in combination with other platforms.

3. Wherever possible, a mobile website built on a standards-based content management system should be at the core of every mobile application project.

4. Digital content should be conceived for cross-platform use and re-use, and developed using the mobile content standards and quality metadata.

5. For preference, mobile projects should use central web services and content should be available through EDAN and/or DAMs.

## Product Development Principles

6. Wherever possible, existing mobile code modules should be reused from the museum's repository: avoid writing new and/or dedicated code and using proprietary or dedicated systems.

7. Whenever possible, make code, tools, best practices and other learning from past projects freely available to others to reuse.

8. For quality and consistency of experience, mobile initiatives should use standard interfaces and include clear, easy routes to other relevant mobile products and platforms.

9. Embed metrics and analytic tools in every mobile product, and include audience research and product evaluation in every mobile project to inform iterative development and ensure quality.

10. Every mobile project or product must include a commercial or other plan for its sustainability and maintenance.

# Notes

1. This chapter is an updated and expanded version of a portion of the essay, "Getting On (not under) the Mobile 2.0 Bus: Emerging issues in the mobile business model" co-written with Allegra Burnette, Peter Samis and Rich Cherry and published in J. Trant and D. Bearman (eds). *Museums and the Web 2011: Proceedings*. Toronto: Archives & Museum Informatics, March 31, 2011. http://conference.archimuse.com/mw2011/papers/getting_on_not_under_the_mobile_20_bus The author would like to thank her collaborators for their contribution to this current version.

2. Comment by Bill Thompson of the BBC during his interview by David Rowan of Wired Magazine as part of the conference, "Mobile for the Cultural Sector," London, 8 March 2011, http://www.camerjam.com/events/mobileculture#agenda~DAYONE-8thMarch consulted 28 March, 2011.

3. In this volume, Peter Samis ("Models and Misnomers for Mobile Production") discusses how these new entrants to the field have exploded the traditional mobile content production model for museums.

4. Loïc Tallon found what may be the earliest audio tour in a museum, the 1952 multilingual audio tour of an exhibition "Vermeer: Real or Fake" at the Stedelijk Museum in Amsterdam: Gescheidenis (History), "Draadloze rondleiding in het Amsterdamse Stedelijke Museum," film clip from Polygoon Hollands Nieuws, July 28, 1952. Details at http://geschiedenis.vpro.nl/artikelen/19265092. Consulted January 30, 2011. Tallon blogged about this discovery at http://musematic.net/2009/05/19/about-that-1952-sedelijk-museum-audio-guide-and-a-certain-willem-sandburg/ (Tallon 2009).

5. Glen Barnes, comment on the online paper, "Getting On (not Under) the Mobile 2.0 Bus: Emerging Issues in the Mobile Business Model | conference.archimuse.com",

http://conference.archimuse.com/mw2011/papers/
getting_on_not_under_the_mobile_20_bus

6. White House, "Open Government Policy", Open Government Initiative,

http://www.whitehouse.gov/open/documents/open-government-directive consulted 28 March 2011.

7. Office of E-Government and IT, Office of Management and Budget, "Data.Gov Concept of Operations", Data.gov http://datagov.ideascale.com consulted 28 March 2011.

8. Koven Smith, "Mobile Experience Design: What's Your Roll-Out Strategy?" in this volume.

9. The trajectory of this partnership, including the branding issues that arose when the museum released its own app and the subsequent decision to temporarily remove the Iconoclash app (pending re-release under a different developer name and branding) has been charted through the Brooklyn Museum's blog; see especially: "Brooklyn Museum API: the iPhone app" 17 April 2009 http://www.brooklynmuseum.org/community/blogosphere/2009/04/17/brooklyn-museum-api-the-iphone-app and "App Store Confusion Necessitates API Changes" 1 December 2010 http://www.brooklynmuseum.org/community/blogosphere/2010/12/01/app-store-confusion-necessitates-api-changes Consulted 28 March 2011.

## References

Anderson, Maxwell L. "Prescriptions for Art Museums in the Decade Ahead," CURATOR: The Museum Journal, Volume 50, Number 1, January 2007.

Kaufman, Jason Edward, "Text $ for Post-fire Phillips Collection." ArtInfo, September 9, 2010. http://blogs.artinfo.com/inview/2010/09/09/text-for-post-fire-phillips-collection. Consulted January 31, 2011.

Burnette, A., et al., "Getting On (not under) the Mobile 2.0 Bus: Emerging issues in the mobile business model," in J. Trant and D. Bearman (eds). *Museums and the Web 2011: Proceedings.* Toronto: Archives & Museum Informatics. Published March 31, 2011. http://conference.archimuse.com/mw2011/papers/getting_on_not_under_the_mobile_20_bus

DaPonte, Jason. Keynote presentation at Tate's 2010 Handheld Conference. http://tatehandheldconference.pbworks.com/w/page/28353070/Jason-DaPonte-Keynote. Consulted January 31, 2011.

Fern, MJ, "Why Apple's New Subscription Model is a Strategic Mistake," fernstrategy blog, February 18, 2011, consulted 28 March 2011 http://www.fernstrategy.com/2011/02/18/why-apple%E2%80%99s-new-subscription-model-is-a-strategic-mistake/

Mobilemarketer 2010. Butcher, Dan, "Red Cross Haiti campaign attracts 3M unique mobile donors." http://www.mobilemarketer.com/cms/news/messaging/7169.html. Consulted January 31, 2011.

Shapiro, Aaron, "The Great App Bubble," Fast Company, August 20, 2010. http://www.fastcompany.com/1684020/the-great-app-bubble. Consulted January 27, 2011.

Samis, Peter, "Models and Misnomers for Mobile Production" in Proctor, Nancy, ed., *Mobile Apps for Museums: The AAM Guide to Planning and Strategy,* Washington, DC: 2011, The AAM Press, American Association of Museums.

Smith, Koven, "Mobile Experience Design: What's Your Roll-Out Strategy?" in Proctor, Nancy, ed., Mobile Apps for Museums: The AAM Guide to Planning and Strategy, Washington, DC: 2011, The AAM Press, American Association of Museums.

Tallon, Loïc. "About that 1952 Sedelijk Museum audio guide, and a certain Willem Sandburg," Musematic, May 19, 2009. http://musematic.net/2009/05/19/about-that-1952-sedelijk-museum-audio-guide-and-a-certain-willem-sandburg/. Consulted January 30, 2011.

| 3 | TED FORBES |
|---|---|
| **Native or Not?** | |
| **Why a Mobile** | |
| **Web App Might Be** | |
| **Right for Your** | |
| **Museum** | |

SINCE 2008, we have seen an explosion of smartphone applications (apps) available from and about museums. A search using the word "museum" in the iTunes store returns literally hundreds of apps for both the iPhone and iPad. Having an app for your institution provides a service on many levels—it's "cool and modern," it provides information to visitors in a transparent manner without being intrusive to the physical gallery space, and it offers institutions a powerful marketing tool. Many museums, boards of directors and web teams have expressed that they feel compelled to "have an app" in order to be up-to-date with the latest technology revolution. But is that a good enough reason to pour time and resources into a mobile app? And do other alternatives provide a better return on the museum's investment?

There are two types of apps that can be developed: "device-native" and "web-based."

Device-native apps are designed to be installed directly on to the mobile device, and are found in Apple's iTunes Store or the Android Market, for example. All of the leading smartphone operating systems provide a market where users can find apps and install them on their phones or tablets.

"Web-based" applications work inside the web browser. Rather than going to an online store to browse, download and install the application, the browser is used to navigate to a website that is optimized for use on the mobile device and offers app functionality.

While device-native apps have the market awareness and have been recognized as being more powerful in terms of technical features and options, web-based mobile applications offer many advantages as well. The Nelson-Atkins Museum of Art in Kansas City has built their mobile strategy around their web-based app, http://naguide.org, which is a mobile website that allows visitors to use any smartphone to access their mobile tour content. The tour makes nearly 300 audio and visual assets accessible for visitors, either during their visit or away from the museum. In order to use staff time and resources most efficiently, the museum's mobile guide links directly to its collections management system, allowing real-time updates to content and data in the tour. Now that visitors can use their own smartphones in the museum, the requirement of maintaining and distributing mobile devices to those who don't have them in the museum has been reduced. The museum has also included the mobile guide in their marketing materials, not only to increase visitor awareness, but also to simplify its use when visiting the museum. There is nothing to download in the iTunes store before you can use the tour: you just open the web browser and navigate to the mobile site's URL.

## Mobile Web vs. Apps: Pros and Cons

Web-based apps offer several advantages when compared to their device-native counterparts. The primary attraction is that a web-based app will work on a greater range of devices, since it is accessed through the mobile web browser, whereas device-native apps only work on the specific device and operating system for which it is designed. If you want your app accessible on the iPhone, Android and Windows mobile platforms, you'll have to develop three different applications. Android is an open-source system and comes in many "flavors." This makes it very labor-intensive to ensure its compatibility across all Android variations. In contrast, browsers are being designed to support standards-based technologies, so a web-based app will work on just about any device.

Web apps also allow institutions to leverage web-based technologies, which generally they already support. These applications are written on the front-end using HTML, CSS and Javascript, all technologies used in website development. On the backend, you'll find a content management system that interacts with a database—again, no different from what museums are already doing with their own websites.

Because the app is accessed through the browser, there is nothing to download, purchase or install. The user simply connects to the website. This is also extremely useful in terms of accessibility outside the museum, as it provides the ability to revisit content from the visitor's own device at a later date. This type of "post-visit enrichment" adds a potentially huge value to an institution and its interaction with visitors.

Another great advantage of the web-based app is that updates and changes to content or design take effect immediately; there is nothing to download. Changes on device-native apps have to be distributed, via iTunes, for instance, and users are responsible for installing the updates on their devices. This can be a time-consuming process, assuming that your users update their apps at all. Likewise, apps and content stored "natively" on devices owned or leased by the museum require staff resources and attention to maintain and update.

Web-based apps can be much cheaper than native apps. Initial development costs for a first generation device-native app can range from $10,000 to $60,000.[1, 2] This is for version 1.0. But then the institution will have ongoing issues of ensuring that the software is current and operates properly as new devices are introduced. For example, when the iPhone 4 was released last year, it included a major software update and a new screen resolution, requiring apps and their images and video to be updated to new, higher resolution versions to support these changes. There will indeed be improvements to processors and operating systems in the future, and basic updates to keep your software relevant are part of the ongoing commitment in time and resources that mobile programs require.

Since web-based apps use existing technologies already installed in your museum, you can likely incorporate a mobile site into your current activities and budgeting processes. In some cases, a web app can be developed by an in-house web team at minimal cost.

Despite the many advantages of web-based apps, there are some downsides to consider as well. Web-based apps live online, so they require Internet connectivity to transmit the content to the device. This could be a serious issue if your institution doesn't have Wi-Fi access with good coverage and bandwidth in the galleries and other areas where you want the web app to be used. Cellular connections may not provide enough bandwidth or stability to support a media-rich web app experience, and visitors on roaming or "pay-as-you-go" data plans will not want to use their own devices unless Wi-Fi is available.

There is a marketing challenge in offering web apps, as well. Apple has an ongoing marketing campaign for iTunes that includes television commercials. People who don't even own an iPhone still understand you can get "an app for that." Web-based apps have to be accessed directly through a URL like a website, so it's up to the institution to do its own marking and search engine optimization to ensure that people find the web app and know that it is available. Web apps can be "bookmarked" to the device's home screen to create the look and feel of a device-native app, but not all users will understand how to do this. That requires more education and marketing from the institution.

---

## Dallas Museum of Art

Drawing from our own experiences with the changing trends of technology and emergence of new devices, the Dallas Museum of Art opted to go with a web-based application to deliver our tour content.

Several years ago, we considered implementing cellphone-based tours in the galleries and throughout the museum. Many museums had adopted this technology as a way of allowing visitors to utilize personal cell phones to access audio tour content. However, the Dallas Museum of Art resides in a building that unfortunately blocks out most phone reception. As we investigated the option of purchasing repeaters to solve this issue, it became apparent that we were looking at a multi-million dollar investment.

Therefore, we decided the smarter solution was to invest in an in-house Wi-Fi network. In 2007 Apple introduced the first iPhone, which profoundly changed the definition of a "smartphone" across the entire industry. Seeing the ensuing frenzy of competition to produce the next "iPhone killer," it became clear that mobile devices were not only going to get cheaper but also more powerful. More importantly, the technology would be changing rapidly. We are an art museum, not a phone provider, so a major concern was that we didn't want to design our entire program around one particular device. Developing an iPhone app, for instance, leaves out Blackberry and Android users. So considering our Wi-Fi commitment and the evolving technology in both the mobile space and the implementation of web standards, we decided the web-based app would give us the most flexibility. The user could use his or her own device. Even for the devices the museum owns and makes available to visitors, the brand is irrelevant and uncommitted. We can change these without affecting the way the content is produced and delivered.

---

### Examining the Needs of Your Own Institution

As you can see, web-based applications come with a few challenges of their own. It is important to ask what kind of app is right for your institution.

In the commercial world, with substantial budget and options, Pizza Hut developed both web-based and device-native apps. After analyzing app use across both platforms, they found that native apps were popular with loyalists, and the mobile web worked well for customer acquisition[3]. So in choosing the right app approach for your institution, the first question is, who is the important target audience?

Is your audience particularly tech-savvy? Most institutions are a mix of tech- and non-tech-oriented people, but does your constituency lean towards one over the other? If your visitors lean in the non-tech-savvy direction and won't bring their own smartphones to the museum, this can mean you have to stock a higher number of devices to hand out, with all the additional costs that entails. And how many visitors do you have in the door, particularly at high-traffic events or openings? Both of these questions will impact the number of devices you'll have to offer on-site in order to provide comprehensive access to your mobile program.

If you are providing your own devices, even in small numbers, you will have to consider staff training issues (or even dedicated staff) for distributing and maintaining the devices. If you have high door traffic, your institution may be required to keep and maintain several hundred devices. What if your visitors have never used an iPod touch and are unfamiliar with the technology? Your visitor services staff might not be equipped to handle long ticket lines while having to stop and provide basic training on how to use the device. You will also have to consider training security guards. When visitors have technical problems, they will not walk all the way back to visitor services; they will more than likely go to the closest museum employee. These are logistical issues to consider that have nothing to do with the web team developing a great app.

Since a web-based application is dependent on an Internet connection, you must also consider your Wi-Fi and bandwidth capabilities. We haven't had any major issues at the DMA, but we have also invested heavily in our wireless connectivity over the last few years. It is important to note that in-gallery Wi-Fi is not possible for all museums, for reasons ranging from cost to having a historically designated building that prevents the staff from making the alterations needed to install Wi-Fi. Smartphones often come with 3G or 4G wireless Internet capabilities, but use of these services can result in expensive roaming fees for international visitors.

So what do you give up in terms of features when choosing a web-based app over a device-native app? Considering both have been developed to run on modern smartphones, really very little. When their app was banned from iTunes for "competing with the phone's native functionality," Google used HTML5 to create a web-based version of their popular Google Voice phone service, successfully circumventing Apple's restrictions.

HTML5 is an emerging web standard that offers many of the features and functions that until now have been possible only in native apps. Device-native apps can take full advantage of access to the core layers of technology the device provides. These include things like a built-in ability to play audio and video, storing data, use of built-in device buttons controls, animations, etc. Web-based applications have more restricted and limited access to these core functions, but HTML5 now includes many of these features (media playback, geo-location, etc.) in its own frameworks and systems; indeed most of these are already supported on modern phone browsers. HTML5 also allows you to create a fall-back if the browser doesn't support a particular feature that you want to build into your web app. Javascript is commonly used to work around certain features that aren't fully supported at this point. It is estimated that HTML5 will be supported across all major browsers by 2014, and it is designed to be backwards compatible, so you can certainly start using it now in the knowledge that it provides fallbacks, meaning if a browser doesn't yet support a feature, you can assign a "backup plan" of either reduced functionality or an alternative feature.[4]

_____

### In Conclusion...

In developing their mobile strategies, it is extremely important that museums consider the full range of mobile platforms available to them, and not limit themselves either first or foremost to device-native apps just because they are "cool and sexy."

Ultimately your primary concerns are user experience and the quality of the content that you offer. It is extremely important not to let your mobile strategy get in the way of this. Technology changes, and visitors don't come to your museum to use iPods. They come for a much deeper experience. It is our job to deliver that.

An iOS app (developed for use on the Apple operating systems) is a tempting solution on many levels: iPhones, iPod Touches and iPads are very popular and give the impression that your institution is using the most current available technology. It is a marketing tool that rides right on the back of the millions of dollars being spent to promote the iTunes store. But pull back and look at this from a 20,000-foot view. What are the costs associated? Do the development costs create a program that is sustainable and able to evolve? What about future devices that haven't been conceived yet? What other devices will be relevant in the coming years? What is necessary to have Wi-Fi in-gallery at your museum? How does this affect the user's experience? Will visitors be frustrated and distracted, or will they find the content useful? These are all important things to consider. A mobile strategy is a serious and ongoing commitment, not a one-off project.

The mobile web can offer museums cost-effective and sustainable solutions for many of their needs, without the two-week-plus wait for Apple to vet the app and possibly even reject it. Nor are web applications limited to creating traditional tour guide experiences. Tours are obviously a priority for most institutions, but there are worlds of possibilities of projects that would encourage participation and interaction with visitors, and the interactivity of the web supports these. I honestly believe we are only limited by our imaginations and the risks we are willing to take.

## Notes

1. "What does it cost to make an iPhone app?" Toy Lounge, 2010, http://www.toylounge.com/howmudoitcot.html (17 September 2010).

2. "How much does it cost to make an app?" London Smartphone, 2009, http://londonsmartphone.wordpress.com/2009/04/08/how-much-does-it-cost-to-make-an-app/ (8 April 2009).

3. Giselle Tsirulnik, "Pizza Hut exec reveals how branded app achieved 2 million downloads,"

The Mobile Xperience, 2010, http://mobilemarketingnyc.blogspot.com/2010/11/pizza-hut-exec-reveals-how-branded-app.html (2 November 2010).

4. If you are interested in reading more of the technical specifications, I highly recommend Mark Pilgram's wonderful web manual, "Dive into HTML5," http://diveintohtml5.org, as well as Erik Wilde's dretblog featuring his wonderful HTML5 Landscape Overview, http://dret.typepad.com/dretblog/html5-api-overview.html.

<table>
<tr><td>

4

**Looking Around vs. Looking Down: Incorporating Mobility into Your Experience Design**

</td><td>

ED RODLEY

</td></tr>
</table>

## Getting Beyond the Tour: The First Questions

Historically, the first mobile museum experiences were tours. From the Stedelijk Museum's radio tours in the 1950s, through the eras of the audiocassette tour, CD tour, PDA tour, and now mobile phone-based tour, the form has remained relatively unchanged. The content has gone from analog to digital, and pictures and video have been added to the narration, but they follow the same model: the visitor goes from location to location and receives content at "stops." Sixty years later, tours are still the dominant type of mobile experience created by museums, according to Pocket-Proof and Learning Times' 2011 Museums and Mobiles Survey. Tours will be offered in some form by 36% of the museums responding to AAM's 2011 Mobile Technology Survey.

However, in the age of the smartphone and the tablet, a mobile museum app doesn't need to be a tour. It can be an interactive book, map, or catalogue. It can be a game, or something entirely different—some new format that takes advantage of the combination of inputs, connectivity and computing power that mobile devices contain. There are some terrific examples of non-tour apps out there: the user-generated soundscapes of *Scapes,* the augmented reality views of historic London in *Streetmuseum,* and the art collecting card games in *Tate Trumps,* among others. All of these apps rise to the challenge of using the capabilities of a mobile device, and don't just treat it like a portable computer for tiny websites or a multimedia playback device. These kinds of experiences hold great potential to deliver compelling content in a manner that connects with audiences, both within and beyond the museum's walls. And reaching a broader audience using these devices is one of the most important reasons to seriously consider developing mobile experiences. Mobile experiences hold the promise of giving museum visitors a new way to deepen their engagement with the institution, while bringing in and (hopefully) retaining new audiences by making the museum more immediate, accessible and relevant.

## Why Mobility?

AAM's mobile technology survey projects that a third of all museums in the United States will introduce *new* mobile technology platforms in 2011, and that smartphone apps will experience the fastest growth. That's a pretty big bandwagon, and the urge to do something because "everybody's doing it!" can be hard to resist. But before you jump on board, you need to be able to answer the fundamental question of "Why make a mobile experience?" In an era of austerity, investing in an emerging platform at the expense of any of the other platforms and projects that might serve your visitors' needs is a big deal and should warrant a carefully considered response.

## Heads Up or Down?

In the museum context, most mobile apps offer two kinds of experiences: immersive, introspective ones that draw the user's attention to the device— "looking down" experiences; and contextualizing ones that direct visitors' attention out into the world—"looking around" experiences. Many combine elements of both.

Mobile games are a great example of looking down experiences. Good ones grab a user's attention and hold it for the duration of the game. Trying playing

*Angry Birds* and doing anything else and you'll understand. Traditional audio tours are classic "looking around" experiences. Ideally, their content is designed to direct your attention outwards, towards the exhibits or sometime towards interaction with your visiting companions. SCVNGR is a great example of an app that combines elements of both: you focus on the device to navigate from location to location, but at the challenge locations you are directed to perform actions that require interaction in the physical space.

In contexts where close observation of an exhibit is the aim of the experience, screens and other "heads-down" experiences have sometimes been considered anathema. But good content and design choices, e.g., showing the detail on the screen and using audio to help visitors find it in the object in front of them, can make the interpretation a scaffold for deeper engagement rather than a distraction from the object. Social and gaming experiences have similarly been held suspect for drawing attention away from the exhibit, but often an exchange or interaction with other visitors can have a more profound learning impact than hours of silent, solitary looking.

The aim should be for the technology to become as invisible a support for the experience as possible. The mobile tours I have tried actually fare worse than their audiocassette and CD ancestors in terms of getting out of the way of the users. They require much more fiddling and time spent looking at the screen to get the next "stop" or interpretive message to play. This rest of this essay seeks to give you ways to avoid these new technology pitfalls and answer these questions for yourself, so you can create compelling, platform-appropriate mobile experiences that go beyond the tour.

## Some Hallmarks of Good Mobile Experiences

The "right" answer to the looking down-looking around question will depend on the design constraints of your project. The most successful mobile apps I've seen to date share many of the following hallmarks.

## They're appropriate for the medium

Whether you're designing a mobile app as part of a larger project, adding a layer of mobile content to an existing experience, or doing something completely mobile, you need to acknowledge and build on how people already use smartphones. People do all kinds of interesting things with these devices. They use them to:

- communicate with other people (voice, text, email)
- listen to audio and watch videos
- access digital information (onboard and streamed)
- play games
- navigate the real world (GPS, AR)
- take and share pictures and video

Chances are, they're doing all these things right now in your museum. When I look at the mobile experiences I've enjoyed the most and gotten the most out of, they are uniformly ones that take advantage of the capabilities of the platform and make it self-evident why someone should use it. They are not just porting an existing experience from one platform onto the new mobile one—like Web 1.0 brochureware websites: they are doing something that can only be done (or can best be done) with this technology. The American Museum of Natural History's Natural History Explorer app has a mapping feature, but it's not just a map. It shows you where you are in real time, something you can't do with a paper map.

## They are relevant first to visitors, not the institution

Be relevant to visitors, and go from there to connect to your own institutional priorities, not the other way around. This means you need to collect data about visitors and not rely just on anecdotal evidence. One of my favorite features of the *Scapes* app was that it relied exclusively on visitors' recorded comments on the artworks and the environment to create the audio interpretation of the

DeCordova Museum's sculpture park. This is a radical approach to guaranteeing the relevance of the mobile app to the visitors by completely eschewing the traditional curatorial interpretive view and voice. Perhaps it was only possible because it was installed as an artwork, not as an audio tour. But perhaps not.

When the American Museum of Natural History launched their Natural History Explorer app, the fanfare was not over the world-class content AMNH provided, but the real-time navigation feature that allowed visitors to navigate the Museum's 46 exhibition halls with some confidence that they would get where they wanted to go. Solving the navigational problem may be less glamorous than an exclusively content-driven app, but in terms of making visitors feel comfortable, I can't think of a better app for AMNH to have launched.

## They encourage authentic visitor feedback

Mobiles are primarily communication platforms, and ignoring that is a terrible oversight. Dialogue is a term that gets used quite a bit in discussions on how museums have to change in order to survive in the modern world. Encouraging participation, designing explicitly participatory experiences, and providing opportunities for visitors to "connect" with museums are popular museum conference session topics. An important step that often seems to get overlooked or taken for granted is "Why do you want visitors to talk to your institution's staff?" It is very hard to have a meaningful conversation of any kind without a shared desire to exchange information, opinions and insights. Museum comment-card boxes the world over are full of feedback that isn't very useful: from "Cool! Loved it!" to "This museum is dumb," many visitor comments are not really actionable items. "Like" buttons can be the same. They're flattering, but don't necessarily lead anywhere. Just as you wouldn't start up a conversation with a stranger without some kind of topic or reason in mind, you shouldn't expect visitors to respond meaningfully unless you provide them a context.

You want a killer app? Invite visitors to take pictures of broken things and send them to your facilities maintenance staff, or tag labels that have typos or out-of-date information. Recruiting visitors to be collaborators, and not just passive consumers of the museum's knowledge and content, will transform your relationship with your visitors faster than any tour.

## They possess a narrative

It's difficult to overstate the importance of narrative. We are a species of storytellers. It seems to be how we make sense of the world. One of the chief mantras at Pixar is "Story is King!" For all of their technological sophistication, they are ruthless about protecting the narrative from being overwhelmed by whatever new technical effect they can create. Everything they do serves the story, because the story is what the audience remembers. Creating a story to tell with your artifacts and experiences is worthwhile because a good narrative can carry so-so content. Good content has a much harder time carrying a so-so narrative. Walking Cinema's *Murder on Beacon Hill* is a good example. The app focuses on a gruesome high-society murder in 1840s Boston. Even though virtually none of the sites associated with the crime still exist, the story is engrossing enough to motivate users to explore the sites picked by the developers. Museums are already repositories of great content, and great narratives. It is a matter of uniting one with the other.

It is important to remember that "story" doesn't necessarily mean a linear narrative of the *"Once upon a time … And they lived happily ever after"* variety. There are many different kinds of narrative structures a museum developer could incorporate in a mobile experience. Non-linear and hypermedia storytelling has become part and parcel of computer game development. Alternate Reality Games (ARGs) like *World Without Oil,* or in the museum sector, Smithsonian's *Ghosts of a Chance* and *Pheon,* have used narrative structures to connect people with each other. Check them out.

*[handwritten margin note: Mystery? Create some kind of a mystery for visitors to solve?]*

## They don't skimp on quality

"Quality" is important in both media format and content design. Good pictures, stereo sound and quality production can make a huge difference in how people react to your app. Part of that is being mindful about how you produce what you put on that tiny screen. Repurposing video meant to be seen on a computer screen or TV can often be problematic, since mobile screens are so much smaller. If you're looking at an iPhone or Android phone, even the latest, greatest models, you're not going to be able to resolve details that you'd be able to see on a larger screen. Watch *Lawrence of Arabia* on a mobile and you'll see what I mean. Picking a content design approach that works on a small screen, like a close-up of an interview subject's head as opposed to a full body shot of the same person, will make your app that much more useable.

## They're free, in some form

One of my favorite things about the current media landscape is that much content has become cheap, or free. I've always hated the "separate fee" model that many audio tours in the past have followed. Doing all that work to build a mobile program, knowing that *at best* 30% of the potential audience is going pay for it, is a real frustration. That said, development—and especially quality content— costs money. Particularly in an emerging market, where the value of the experience is still unknown, being able at least to recoup some costs may mean the difference between starting a project or not. The more you can lower the price bar to initial entry to the experience, the more likely that people will try it.

Giving the mobile program away in some form is the best way, particularly in the smartphone arena where free apps are so prevalent. The madness of people balking at paying 99¢ for an app on a phone they spent $300 to buy and $50+ per month to use is a topic for another day, but it's the reality of the present smartphone landscape and needs to be acknowledged. Getting people to take the time to download and launch your app is the biggest hurdle you're likely to face. If you get that far and face the problem of how to get them to use it twice, you're already ahead of the game.

### Getting Started

Probably the best thing you can do before you start planning your own mobile app is to find and use as many compelling mobile apps as you can. Think about why you like the ones you like, as a *user*, not as a developer. Be shameless in building on the work that others have done. The web is full of great repositories of museum mobile experience. Do some real research on your intended audience. And most importantly, don't wait for "the market to stabilize" and resolve the uncertainties that swarm around the mobile sector. iOS or Android? Native app or web app? HTML5? One of the most reassuring projections I took away from the 2010 Tate Handheld Conference was the consensus that the idea that we were heading to a promised land where the technology would no longer be in flux was unrealistic. It will always be changing; there will always be some new paradigm-upsetting product or service in the wings. Ted Forbes, multimedia producer at the Dallas Museum of Art, summed up his mobile strategy at DMA better than any I've heard, so I'll steal it: "Do it now. Do what you can. Do it better tomorrow."

Good luck!

# 5
# Models and
# Misnomers for
# Mobile
# Production

PETER SAMIS

ONCE upon a time there was Audio Tour Company A, and Audio Tour Company AA, and they pretty much ruled the roost. Their business model was sound: Help museums communicate with their visitors by telling stories that connect people with specific objects and broader themes. This was at a time when professional audio expertise—writing for the ear rather than the page; directing and recording voice talent; cutting in expert voices; and adding a pinch of sound design panache—was especially hard to come by, long before any museums thought of themselves as media producers. Why this was way back ... five years ago!

These companies thrived on the interpretation of two types of exhibitions: blockbuster, popular crowd-pleasers on which they made their nut; and permanent collection shows, which were loss-leaders and cost their museum clients real money to make. (How many stops did you say you wanted? And would you like a family tour with that?)

Companies such as our two *ur*-entities also duplicated the tapes (remember the Walkman?) or .mp3s to whole fleets of portable players in a progression of shapes and sizes, usually with excellent quality control. They translated and re-recorded scripts in multiple languages. They tried their hand at inventing new, ever more compact, convenient, and rugged players, which they leased to museums and which the museums rented out in turn to their visitors—often with the help of trained staff supplied by the companies. By and large, it was a harmonious model, and all seemed right with the world.

Then the world changed. Thanks to software and hardware advances, it became far easier to produce audio and video, and a new generation came of age and they were called "Digital Natives." There where their parents feared to tread, they knew no fear.

Museums, for their part, began to pay more than lip-service to the public's "right to know." In some cases—though by no means all—an ethos of universal access to interpretation emerged: Interpretation as a right, not a privilege with a price tag on it. A perfect storm emerged:

•The profit motive for audio interpretation dwindled;

•The desire to offer audio interpretation for more exhibitions grew, including the so-called "difficult," unpopular ones that for-profit audio tour companies had typically neglected (not a fit for their profit model);

•The Digital Native generation came of age and began to work in museums.

The old model shaken, empires buckled, and we find ourselves today in a dramatically altered landscape set against the debris of old devices now rendered obsolete: diasporas of audio engineers, creative producers, and script writers; new DIY software vendors; new consumer platforms ranging from iPods to smart phones; and lots of possible permutations for museums eager to move forward with mobile interpretation for their visitors.

Some museums opt to have their curators and content experts "phone in" their commentaries and publish them to visitors' cell phones. Others conduct and edit interviews, script narratives, and produce their own tours, with or without outside help, and then publish them using any of a variety of mobile content management software systems now on the market. They make them available on devices they lend to visitors and/or visitors' own devices via the Web or an app store. Others still prefer to delegate the whole kit and caboodle—content production, publishing, hardware provisioning, staffing and distribution—to a single outside company for a turnkey solution, as in the old days.

All solutions are possible, and different situations may call for different responses, even in the same museum. Consider the grid below.

| | Script Development | Media Production | Publishing to Devices | Hardware Provisioning | Mktg, Sales & Distribution | Analysis & Evaluation |
|---|---|---|---|---|---|---|
| MUSEUM ALONE | | | | | | |
| MUSEUM WITH VENDOR | | | | | | |
| COMPLETELY OUTSOURCED | | | | | | |

There is no "one-size-fits-all" answer for how to approach your own project. Any given square in the grid can be a "right" answer, and your museum can move from one row to another—from working alone to outsourcing—as it proceeds through the phases of your project. For a given task, museums can:

• choose to do the work themselves

• collaborate with an outside vendor in such a way as to build their in-house expertise

• completely outsource the task

To determine which model is best suited to your museum, do an assessment of your in-house talents, skills, and capacities—and your potential outside partners. Think outside current job descriptions. *Talents* are things at which someone on your staff excels (storytelling, for example, or digital media production). *Skills* are things you can do well. And *capacities* are something you or someone on your team *could* do—or learn to do—if required. *Partners* are your network of outside resources: people, companies, or collaborating institutions you could call upon to help. Look at your untapped talent pool—or consider hiring or training people to do things that were never previously part of the museum staff skill-set.

You may decide it is more important to devote staff time to building expertise and interpretive resources around the permanent collection, and outsource the temporary exhibitions. Or you may decide that you want to publish to visitors' own devices, thereby sparing your museum the expense of purchasing and maintaining hardware and staffing distribution desks. But be forewarned that by doing so you risk perpetuating the Digital Divide, making a situation where only those who can afford it (or have the chosen device) have access to interpretation. Many museums keep a fleet of devices on-site—from as few as 20 to 200 or more—which they offer to visitors to check out with the deposit of some form of security.

You may also decide that the solution you develop for "business as usual" falls short in the event of a blockbuster, for which you will need many hundreds of devices and one or more dedicated distribution desks with their own trained complement of staff. In today's transitional environment, the old rules no longer apply. You can put out an RFP to several mobile tour companies for the servicing of that specific show and retain your in-house solution for less encumbered times.

We're entering a mix-and-match media world. Keep copyright to any content you develop and make sure it is available to you in file formats compatible with future re-purposing to other platforms. Assess (and grow) your capacities, skills and talents, choose your partners to complement your in-house strengths, and be prepared to adapt to new opportunities as they unfold. In these times of technological change, it's the mission, not the means, that will be your fixed target.

Margriet Schavemaker

6

**Is Augmented Reality the Ultimate Museum App? Some Strategic Considerations**

IN the past year, the innovative forms of augmented reality (AR) appearing on smartphones have proven to be exciting playgrounds for curators and museum educators. These AR tools offer users the possibility to deploy their phones as pocket-sized screens through which surrounding spaces become the stage for endless extra layers of information. This visual collision of the real and the virtual—made possible by using GPS and a compass—could culminate in what we have seen in movies like *Minority Report* (2002), where Tom Cruise physically navigates through 3D data: a seamless interface between the body, the virtual and the real.

Currently, however, AR technology (Layar or Junaio, for instance) is still a kind of experimental medium, as yet lacking the total immersion that science fiction promises. Moreover, its mediation through a tiny handheld screen poses several challenges to augmented storytelling. What, then, does this contemporary form of AR have to offer the museum today? Why would a museum want to develop augmented reality tours? What kind of user experience does it entail? Is it, in this day and age, the ultimate app? These questions will be addressed here by taking a closer look at the experiences of the Stedelijk Museum's AR project, ARtours, which explores a number of augmented reality applications in order to experiment with these new platforms in different contexts and with different kinds of art.[1]

*Lieux de mémoire, space hacking & artistic platform*

Taking a closer look at the deployment of AR by museums, it seems that the attraction of this new medium is often found in the act of returning cultural heritage to the streets where it was originally produced and/or that it depicts. As the apps of the Powerhouse Museum[2] and the London Museum[3] effectively illustrate, AR allows users to see photographs on their smartphones of old city views overlaid on the places that they were shot. Comparing a "real" contemporary with an "augmented" older view offers a moment of reflection on history, modernization and change.

The Netherlands Architecture Institute (NAI) included even more time dimensions in its unequaled application (UAR)[4], as visitors are not only treated to former architectural drawings of the locations where one is positioned, but also to unrealized designs and future projects. The strategy, however, remains the same: using AR as a medium to layer the urban realm with a museological collection in order to compare its current outlook with that of other times and ages. In a sense, it is using AR as a form of what Pierre Nora would describe as *lieux de mémoire.*

For a modern and contemporary art museum like the Stedelijk Museum in Amsterdam, this strategy for AR deployment is relevant in that the word "Stedelijk" means "municipal," and parts of the collection are produced by or related to the Amsterdam cityscape. However, layering the streets and canals with these local artworks has certainly not been the main reason for investing so much energy in the development of *ARtours.* First and foremost, the museum is known for its extensive international collection of art, photography and design, which itself asks for a different curatorial approach and visitor experience. Secondly, the Stedelijk Museum has been closed since 2004 due to a renovation of its original building and construction of a new wing. AR was therefore primarily embraced because of the possibilities it offers for exhibiting the collection, as the museum has lacked an analog venue in which to do so. In other words, in addition to *lieux de mémoire,* the Stedelijk opted for space *hacking,* a strategy in which augmented reality is used to present the collection in spaces with which the art has no relation whatsoever, but are used simply as a new stage.

We experimented with this strategy in the *ARtours* project entitled "ARtotheque."[5] The idea is simple: the Stedelijk Museum holds thousands of artworks in its collection, so why not lend copies to the general public via the medium of augmented reality so that people can place the artworks wherever they choose? The project location can be anywhere; we experimented at Lowlands (a Dutch music and arts festival with 50,000 visitors) and at the innovators' festival, PICNIC. Participation was relatively simple: the visitor could choose an artwork from a selection of 160 masterpieces, all printed on A4 cards, scan the QR code on the card and thus activate the "ARtotheque" (art loan) layer on the Layar platform. The visitor could then choose a position for the artwork, hang it and share it with all other works in the public "ARtotheque" layer.

As the Stedelijk Museum is also known for its contemporary art projects, another utilization of AR appeared relevant: augmented reality as an *artistic platform*. In the *ARtours* project entitled "Me at the Museum Square," ARtours experimented with this strategy by asking students from various Dutch art schools to design an augmented reality artwork to be virtually manifested on the large square adjacent to the museum.[6] Stedelijk curators made a selection of the most promising ideas, and together with students from the University of Amsterdam and the School for Interactive Media (project Medialab), the 3D "Artworks" were realized. Besides helping the project to get a better grip on the possibilities of Layar and the practical problems AR applications pose to users (too much sunlight, battery consumption, etc.), another result of this project was the fact that several of the created artworks reflected upon the new medium.[7] For instance, in one work audience members could virtually augment themselves with auras in various colors, which derives from the artist's idea that AR is, similar to auras, visible for some and not for others. Another artist placed a springboard next to the small pond on the museum square. The title of the work, "The most fun you will never have," addressed the fact that, in augmented reality, the virtual is colliding with the real but not transforming into the real (in a material sense). It is this kind of self-reflexivity that helps us in coming to terms with AR's cultural significance.

---

## Let's go inside

In the summer of 2010 the Stedelijk Museum got the old part of its building back. The renovation was almost finished and, although the additional wing was not yet ready, the museum could make a start with temporary exhibitions and public programs. For the *ARtours* project, this signified an interesting strategy shift to bring AR out of the streets and into the white cube.

As early as 2002 media theorist Lev Manovich claimed that, with augmented space,

> …museums and galleries as a whole could use their own unique asset—a physical space—to encourage the development of distinct new spatial forms of art and new spatial forms of the moving image. In this way, they can take a lead in testing out one part of the augmented space future…

> Having stepped outside the picture frame into the white cube walls, floor, and the whole space, artists and curators should feel at home taking yet another step: treating this space as layers of data. This does not mean that the physical space becomes irrelevant; on the contrary, as the practice of [Janet] Cardiff…shows, it is through the interaction of the physical space and the data that some of the most amazing art of our time is being created.[8]

The *ARtours* project selected for its first indoor *AR(t)* project artist Jan Rothuizen, known for his hand-drawn maps on paper.[9] In the AR application Rothuizen's drawings are virtually appended to the spaces of the building to which they refer. Using a smartphone you can open the tour and follow Rothuizen's childhood memories of the museum throughout the gallery spaces.

Also included are his references to the Stedelijk's renowned history and close observations of the institution made while spending a night in the building.

The result is a layering of the real with virtual information, bringing the objective outer world of material spaces into collision with the subjective inner world of conceptual memories and storytelling: a mapping of the museum inside the museum that echoes the psychogeographical maps produced in the 1960s by the French Situationists.

Of course the move from outside AR to inside was not that easy, as current technology (Layar) relies on GPS to attach the virtual to the real. GPS has difficulty in distinguishing vertical levels inside a building; thus additional interfaces are needed to delineate one's location inside the building. Since these methods of interface have not been perfected yet, we are pleased that AR providers are exploring new solutions to the problems of bringing the technology indoors. The *ARtours* project will experiment with these in the near future in collaboration with Fluxus artist Willem de Ridder, who is working with us on one of his "Secret Exhibitions" in AR. Moreover, we are exploring possibilities of bringing a selection of the Stedelijk Museum's famous exhibitions back into the building by means of AR, re-using the museum archives and documentary material.

### Innovation & collaboration

Besides all these more practical and media-related strategies that readily illustrate how and why a museum might use smartphone-based augmented reality, there are more overarching reasons as well, of which "innovation of audience participation" seems the most pivotal one. For the Stedelijk Museum, this seems to fit a long-established tradition: the museum is said to be the first in the world to have created "audio tours," in 1952.[10] Of course the radio broadcast technology used in that time was far from perfect and the experience was almost identical to a conventional guided tour (for instance, people were bound by the tour's time constraints and were not free to move around, being required to follow a linear story). However, as specialist in the field Loïc Tallon rightly makes known, this was not the point. What mattered most was that the audio tours of 1952 were launched by the Stedelijk at the same time that the ICOM conference was held in Amsterdam that year. Consequently, the entire museum world took notice of this new development and many immediately started to develop similar systems. Therefore Tallon concludes that

> Above all, I believe that it was the innovation and potential embodied within the audio guide that best explains why the Stedelijk Museum 'invented' it. Whilst one could claim that what was achieved by the system could have been achieved through trained docents, this is too narrow a perspective. After all, this innovation went on to spawn what was arguably the most successful museum technology of the 20th century, and one of the most exciting of the early 21st century.[11] In 2011, "innovation and potential" also seems to be the driving force for augmented reality applications. It is not about offering the most perfect technological solution and radical new user experiences. Moreover, it is often hard to define differences with respect to existing multimedia tours. However, the potential for bridging the gap between the virtual and the real world in a single visual interface is a dream shared by many and thus a great stimulus for future innovation.

Innovation can only exist through collaboration. In 1952 the Stedelijk Museum created its audio tours with the renowned Dutch enterprise Philips. At present the Stedelijk works with several technological and design partners, such as Fabrique, 7scenes/De Waag, Tabworldmedia and Layar. Collaborations with educational partners (University of Amsterdam, Hogeschool van Amsterdam, art schools) and cultural organizations (Tate, Virtual Platform, ISEA, kennisland) also exist. These partners should not only receive full credit for the *ARtours* project, but should also be thanked for the innumerable innovations inside the Stedelijk organization they have triggered thus far and will continue to do so in the future: from fundamental changes in museum technology (ubiquitous Wi-Fi access) to new takes on copyright issues; from changes in media

awareness and the programming of our educational and curatorial departments to new policies on the future of audio tours in the museum; and so on. For a museum reinventing itself in the 21st century, this is invaluable, and leads to the idea that a museum should always incorporate at least one innovative project like *ARtours* every other year.

### Paratouring

Can we already draw some conclusions about the outcome of the first 1.5 years of the *ARtours* project? Findings that may help other museums to decide whether augmented reality can be their ultimate app? Insights that may fuel debate on the future of mobile technology in the museum?

Inspired by the "un-conference" concept, museum professionals at the industry conference Museums and the Web and elsewhere have discussed for the past couple of years the "untour," referring to the manifold possibilities in our current 2.0/3.0 phase where mobile tours can go beyond the traditional audio tour format.[12]

The *ARtours* project defines another interesting development in the usage of mobile media inside the museum: the "paratour." The term "para" refers to the extra information that normally accompanies the core text of a publication: the introduction, conclusion, notes and additional literature, often provided by

the editor, which are collectively referred to as "paratext." They are the discursive elements that frame the text, positioning it through an extra layer of information.

Of course the traditional audio tour can itself be considered a "paratext," as it frames art with an auxiliary text. However, the *ARtours* project indicates that innovative museum tours, like augmented reality applications, become especially significant by way of extra communication tools and additional layers of information. Significantly, the tours elicit communication among the users. In order to use an AR tour, generally one has to join forces, as not everyone possesses the appropriate smartphone, the user interface is still challenging for some, data traffic is not equivalent for all telecom providers, using the app tends to drain batteries quickly, etc. This turns the AR tour into a social event, something the Stedelijk Museum facilitates by organizing a public program and opening event every time a new project is launched.[13] This form of "paratouring" among users exists not only in the analog world, but extends into the virtual one as well via social networking services like Facebook and Twitter. In addition, the *ARtours* project has opened the eyes of the museum to a ceaseless flow of professional "paratouring" by museum and other mobile technology experts. The innovative mobile museum tour has an amazing, extended lifespan mediated through videos, PowerPoint presentations, lectures, Twitter feeds, blogs, conferences, roundtable discussions, expert meetings, wikis and remarkable press coverage. It may even be the case that the ARtours project has more followers on Twitter and via our blog than people who have actually experienced the *ARtours* themselves.

Of course one can denounce "paratouring"—or, in terms of AR, "pARa-touring"—as a distraction from what the tour is really about, namely, mediating knowledge and enhancing visitor experience both inside and outside the museum. This is a risk, and we should take care that it does not obstruct the actual encounter with the museum, collection or exhibition. Still, we cherish the fact that a museum that has been in hiatus for over seven years is suddenly back in the spotlight![14] If this can happen in the world of mobile media, why not in other fields as well?

## Concluding remarks

If we now return to the central question of this discussion—'Is augmented reality the ultimate museum app?'— we must conclude that, at first sight, it certainly is not: the technology is experimental, the user interface problematic, and we are as yet very far from the ideal future of total immersion and seamless interfaces (as visualized in movies like *Minority Report*).

On the other hand, we have seen that AR can be significant for museums in many ways, both outside and inside the museum, as it:

- offers interesting collisions between virtual (digitized) heritage and real (analog) space;

- provides a new platform for artistic experimentation;

- is a perfect medium for museum innovation and collaboration; and,

- generates enormous amounts of communication, interpretation and contextualization (the so-called "paratouring").

For the Stedelijk Museum, in its current "temporary" phase within and without its building and in the process of reinventing its institutional identity, AR has proven to be the ultimate app! For other museums, the best recommendation may be to consider all relevant strategies … and then engage in it anyway.

## Notes

1. For a more practical and detailed relay of the *ARtours* project, please see our paper "Museums and the Web 2011" http://conference.archimuse.com/mw2011/papers/augmented_reality_ and_the_museum_experience an www.artours.nl

2. http://www.powerhousemuseum.com/dmsblog/index.php/2010/04/16/new-version-of-powerhouse-museum-in-layar-augmented-reality-browsing-of-museum-photos-around-sydney/

3. http://www.museumoflondon.org.uk/English/VisitUs/You-are-here/

4. http://www.nai.nl/tentoonstellingen/3d_architectuurapplicatie

5. http://vimeo.com/15191266

6. For more information on "Me at the Museumplein," see http://ikophetmuseumplein.nl/

7. More detailed information about these projects can be found in http://conference.archimuse. com/mw2011/papers/augmented_reality_and_the_museum_experience

Also take a look at http://vimeo.com/15191542

8. Lev Manovich, *The Poetics of Augmented Space,* 2002 (revised 2004)

9. http://vimeo.com/18092710

10. http://www.openbeelden.nl/media/22823

11. Loïc Tallon http://musematic.net/2009/05/19/about-that-1952-sedelijk-museum-audio-guide-and-a-certain-willem-sandburg/. Of course one should also acknowledge the fact that the point of this tour, and most of the early audio tours, was to provide multi-lingual interpretation where it was not possible to have multi-lingual live guides at all times – so the technology did fill a need the human staff couldn't meet.

12. See, for instance, Nancy Proctor's workshop on the topic of the "Mobile Untour" at MW2011: http://conference.archimuse.com/mw2011/workshop/mobile_untours and notes from the "Un-tour Unconference" session, Museums and the Web 2010. http://conference. archimuse.com/forum/untour_unconference_session

13. For example, an evening program around gaming ("Play it!"...) during which the Stedelijk AR(t) project Jan Rothuizen was launched. http://www.stedelijk.nl/en/now-at-the-stedelijk/ lectures/agenda-lectures-and-events/do-it-play-it

14. See, for instance, the article "Art Gets Unmasked in the Palm of Your Hand" that appeared in the *New York Times* and in which the Stedelijk Museum is one of the major cases next to major museums like MoMa. http://www.nytimes.com/2010/12/02/arts/02iht-rartsmart. html?_r=2&emc=eta1

7

**Mobile Content
Strategies for
Content
Sharing and
Long-Term
Sustainability**

ROBERT STEIN

---

## Sustainable Mobile Content

The 2010 Horizon Report for Museums highlights "mobiles" as one of two technology trends on the near-term horizon, noting that "Mobile technology has developed at a staggering pace over the last few years, and today affords many more opportunities for museums..." (Johnson, 2010) The recent explosion of mobile technology as an important way for museums to distribute content is undeniable. Dozens of new tools and companies have emerged in the past 24 months to address the needs of museums that are planning, producing and launching new mobile experiences. A recent Pew Internet survey indicates that 40% of American adults had access to the Internet from a mobile phone in 2010 (Smith, 2010), and studies from Gartner suggest that by 2013 mobile phones will overtake PCs as the most common method for accessing the Internet worldwide. (Gartner, 2010)

With four billion mobile phone subscribers worldwide, it's clear that mobile devices and content will be an important means of access for museum visitors today and in the future. More recent anecdotal evidence suggests that these trends have continued to accelerate, and that an increasing number of museums are contemplating how they might deliver content via mobile devices. The Museums and Mobile Surveys 2011 indicates that over half of large museums (annual attendance of more than 50,000) already have mobile experiences, and almost 70% of all museums say that their institution will "definitely" have in-house mobile content development within the next five years. (Tallon, 2011)

For museums, however, the relationship between museum content and technical change has always been challenging, given the dramatically different time-scales of the two disciplines. A museum's primary "natural resource" is the content it produces in support of the concepts, collections, and programs that are the source of its mission. Museum collections evolve slowly over many decades, and the concepts and programming created to support the mission of museums are adapted continually, being more an evolutionary optimization of a consistent set of principled goals. This means that most museum content will remain relevant for many years after its creation.

Technology, on the other hand is defined by change. The well-known Moore's Law states that the density of transistors on integrated circuits will double every 18 months. Applied to the rate of change in technology hardware, Moore's law has proven to be an accurate predictor of technical innovation since the 1960s. In the last few years however, it seems that software innovation is out-pacing even this dramatic prediction. A recent *New York Times* article highlights the fact that innovations in software architectures and algorithms have recently trumped even the staggering pace of Moore's Law for hardware innovation. (Lohr, 2011)

This defines a critical issue that is integral to understanding the relation-ship between museums and technology. How can museums flexibly adapt to the rapid changes of technical innovation while leveraging a body of content and collections that change at a comparatively glacial pace? Can museums create and maintain building blocks of content infrastructure that will last longer than any particular iteration of technology platforms?

The creation of open-software tools and standards for mobile tours and experiences that can be shared and referenced by museums and vendors would offer an effective way to answer many of these questions, and would provide a mechanism to ensure that content created today could be easily re-purposed and adapted to future generations of mobile platforms. Building consensus among museums and vendors for a description of mobile content, and building tools to aid in the adoption of this platform, are necessary steps to achieve the goals of content sustainability and cross-collection sharing that museums desire. A successful solution of this kind would provide a way to integrate and interoperate between a number of content-creation systems and mobile interfaces, allowing both vendor-provided and custom-developed application software to use the same set of content.

The key element in achieving such compatibility among mobile platforms is the existence of a specification—a common language—describing mobile content and the experiences they provide. In the summer of 2009, the Indianapolis Museum of Art (IMA) proposed a simple draft specification called TourML (pronounced *tûrmoil*) (Stein, 2009), which offers a working, but preliminary, example of what such a common language might look like.

In order to solicit a high level of input from the community, Robert Stein (IMA) and Nancy Proctor (Smithsonian Institution) organized several free community workshops, inviting museum staff members, academics and mobile vendors to join a preliminary effort to formulate just such a standard. In all, nearly 100 members of the museum community have played a significant role in these workshops, resulting in multiple subsequent revisions to the TourML specification. Notes from all meetings are available from the Museum Mobile Wiki, (Stein and Proctor, 2010) and the resulting TourML specification is available under an open-source license from the project's Google Code Website. (Moad and Stein, 2009) A more complete discussion of the TourML specification was documented in a recent paper. (Stein and Proctor, 2011)

## Putting TourML to Work

Well-defined content specifications like TourML are particularly well suited to function as an interchange format or middleware between authoring tools and presentation tools. Figure 1 shows a proposed use for TourML as an intermediate

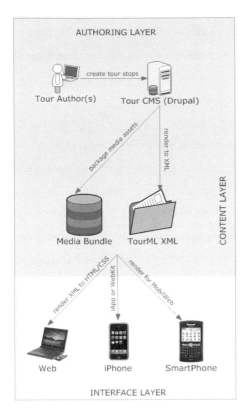

AUTHORING LAYER

Tour Author(s)    Tour CMS (Drupal)

create tour stops

package media assets

render to XML

CONTENT LAYER

Media Bundle    TourML XML

render XML to HTML/CSS

iApp or WebKit

render for MobiWeb

Web    iPhone    SmartPhone

INTERFACE LAYER

Figure 1. TourML used as an interchange format

layer in the publishing workflow for mobile tours. In this scheme, tour authors create content in museum-specific content management systems with support for TourML (i.e., the open-source content management system, Drupal). The content management system can then re-write that content as a TourML document. Both the document and all media assets needed for the tour can be bundled together in a single, platform-neutral package. Web applications or device native apps can be easily created to read the TourML document and access these media assets. Since the TourML document is platform agnostic, the same document can be used for apps on several different kinds of devices.

As a practical example of how TourML might be used, the Indianapolis Museum of Art released an open-source mobile tool called TAP in 2009. (Moad and Stein, 2009) TAP provides mobile tour authoring tools based on a Drupal CMS and publishes the tours as mobile apps for the web and iPod Touch/iPhone (Figure 1). As it builds mobile tours, TAP automatically encodes content elements using the TourML standard, so that whole tours can be easily exported from TAP to other platforms or future authoring systems. The goal is for at least 80% of a tour to be able to move directly across platforms, thanks to the TourML schema, minimizing the amount of human intervention required to customize the tour for its new environment.

Since its release, TAP has been used by the IMA to provide five

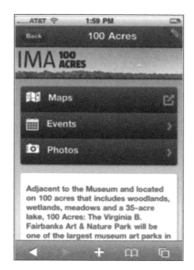

Figure 2. Sample interfaces for TAP including mobile web and iPod versions.

exhibition-related mobile tours and one outdoor mobile tour highlighting the museum's Art and Nature Park. Figure 2 shows a few example screens from selected mobile tours. Since its release, TAP has been successfully downloaded and used by several other museums including the Museum of Fine Arts, Boston (Fleming and Kochis, 2011), and the Balboa Park Online Collaborative. (Sully, 2011)

---

## The Role of Museums and Commercial Partners

The prevalence of software vendors offering mobile products, and the lack of technical expertise on the part of many museums, leads to an important and nuanced relationship between the museum and vendor communities concerning the preservation of museum mobile content. The nature of commercial competition often makes it impractical for vendors to lead the charge for portability and preservation. In addition, without a consensus in the museum community regarding content definitions, tools that ensure content portability and preservation are all but impossible. Previous successes in defining content specifications and standards, such as those supporting collection metadata (LIDO, CDWA Lite, and Dublin Core), have been led by the content producers. In short, the onus of consensus and collaboration around content standards falls squarely on the shoulders of the museum community. But the importance of a healthy collaboration with a community of reliable vendors cannot be overstated. It is prudent to recognize the fact that museums cannot hope to keep pace with technical change without the assistance of commercial vendors who specialize in particular areas. Doing so allows museums to take advantage of the targeted capacities of the most advanced mobile products, while maintaining control of the creation and preservation of its content—a core priority of the museum. Partnership with commercial vendors to encourage adoption of standards, and ensuring that these standards help to enhance the vendor's business, is the only sure way to secure the viability of such an effort.

Through a ratification of the TourML specification or another similar effort, museums can spearhead the adoption of this specification by the vendor community. In explorations of the feasibility of this effort, many commercial vendors have been very receptive to TourML as a potential specification for mobile content, and have been actively engaged in the mobile workshops that have been held. Some vendors have already integrated early support for the draft TourML specification into their products. Ideally, a healthy relationship and collaboration between museums and the vendor community in this process will result in a viable and sustainable specification that can truly produce the benefits we seek.

## References

Fleming, Jenna, Kochis, Jesse, Getchell, Phil. (2011) "Launching the MFA Multimedia Guide: Lessons Learned." Museums and the Web 2011, Philadelphia, Pa. April 2011.

Gartner. (2010) "Gartner Highlights Key Predictions for IT Organizations and Users in 2010 and Beyond." January 13, 2010. Consulted January 27, 2011. http://www.gartner.com/it/page.jsp?id=1278413

Johnson, Laurence F., Levine, Alan, Smith, Rachel S. and Witchy, Holly. (2010) "Horizon Report: Museum Edition." Austin, TX: The New Media Consortium, 2010. http://www.nmc.org/pdf/2010-Horizon-Report-Museum.pdf

Lohr, Steve. «Software Progress Beats Moore›s Law" - NYTimes. com. Technology - Bits Blog - NYTimes.com. 7 May 2011. Web. 10 Mar. 2011. <http://bits.blogs.nytimes.com/2011/03/07/software-progress-beats-moores-law/>.

Moad, Charles W., Stein, Robert J. (2009) TAP-Tours, Google Code Project Site. 2009. Consulted January 27, 2011. http://code.google.com/p/tap-tours/

Smith, Aaron. (2010) "Pew Internet & American Life: Mobile Access 2010." July 7, 2010. Consulted January 27, 2011. http://www.pewinternet.org/Reports/2010/Mobile-Access-2010.aspx

Stein, Robert J. (2009) TourML (In Progress). 2009. Consulted January 27, 2011. http://wiki.museummobile.info/museums-to-go/products-services/tourml

Stein, Robert J., Proctor, Nancy. (2010) Museum Mobile Wiki: Standards. 2010. Consulted January 27, 2011. http://wiki.museummobile.info/standards

Stein, Robert J., Proctor, Nancy. (2011) "TourML: An Emerging Specification for Museum Mobile Experiences." Museums and the Web 2011, Philadelphia, Pa., April 2011.

Sully, Perian. «IPod/iPhone Mobile Tours for Everyone | Balboa Park Online Collaborative.» Balboa Park Online Collaborative | Museum Innovation Through Collaboration. 9 Mar., 2011. Web. 14 Mar. 2011. <http://www.bpoc. org/node/10602>.

Tallon, Loïc. (2011) "Museums & Mobile Survey 2011." January 2011. Consulted January 27, 2011. http://www.museums-mobile.org/survey/

<table>
<tr><td>8<br><br>**Understanding**<br>**Adoption of**<br>**Mobile**<br>**Technology**<br>**within Museums**</td><td>Kate Haley Goldman</td></tr>
</table>

EACH week brings across my desk a fresh set of mobile market studies indicating how the proliferation of smartphones continues at a dramatic pace, web access is more and more mobile, phones have changed teen culture, phones have changed the culture for the rest of us, and smartphone domination is in our near future. Having established near ubiquity of phones in general, most studies foresee a continued acceleration of the adoption of the smartphone. The adoption acceleration has extended to museums, and in recent years is starting to become less talk and more actual products on the floor. Multiple recent studies have tackled the institutional perspective of mobile adoption and barriers to implementation. Among the most relevant for the museum field are the 2011 studies sponsored by AAM and Pocket Proof/Learning Times. This chapter will interpret the findings of these studies and frame questions for future studies.

The study by Fusion Analytics for AAM (supported by Guide by Cell) focused on the state of current and future institutional incorporation of mobile capabilities. The sample was somewhat larger than the study by Pocket Proof, focused primarily on the United States, and had a geographically wide and institutionally broad distribution. One key finding was that under half of American institutions within the study currently offer mobile interpretation. For those institutions that did not yet have mobile, the primary reasons were lack of budget, staff time and other resources. Most non-mobile institutions did not express a concern that they did not want their visitors using mobile devices, but they were concerned about visitor interest. Over one-third of museums without mobile products listed lack of visitor demand as why they did not offer such products. By comparison, the study from Pocket-Proof and Learning Times asked museum professionals about their attitudes towards the creation, implementation, and maintenance of mobile applications, segmenting the results by the differing challenges for those institutions who already have a mobile interpretation product compared to those who are planning to, but do not yet have such as product.

Both studies are admirable in their effort to provide baseline data and context for the proliferation of museum mobile projects. Looking at their data, it would seem that the vast majority of museums are currently working on such projects. And while that generalization may in fact be true, unfortunately we can't be sure from the data presented within these studies.

Generalizability is a tricky proposition with research and evaluation studies. Fundamentally, research is judged on its reliability and validity. Reliability is defined as the consistency or stability of a measure from one test to the next. An accurate oven thermometer is reliable, measuring 350 degrees in the same fashion each time. Validity is the overall term used to describe whether a measure accurately measures what it is supposed to measure. An accurate oven thermometer might be consistent in temperature measurement but does not measure whether the food is hot enough. Oven temperature is not a valid measure of food temperature. It is possible to have reliable but not valid results: results that are repeatable in multiple testing but still do not measure the appropriate underlying construct. The oven might be consistent in the temperature measurement, but it still does not represent a valid measure for the temperature of the food.

Reliability is influenced by the design of the questionnaire, but most profoundly by the sampling within a project. The Pocket Proof study uses a non-probability sampling: respondents were recruited via list-serves, social media and other outreach. While the sample sizes are large; there is no certainty that the population sampled is representative of the overall population of museums in the United States; indeed they are likely to be museums with strong connections to social media, and by extension, use of technology within the museum. This may bias the results in terms of the numbers who have mobile projects or are considering them, or bias in other ways, such as size of institution, type of mobile project attempted, etc.

The AAM study had a well-defined sampling frame: AAM member institutions and individual members. While there is likelihood of some amount of nonrandom measurement error, the issue of most concern is the response rate. Response rates are notoriously difficult on web-based surveys, even for surveys such as this one with incentives offered and a dedicated (but busy) membership. At what point can a response rate be deemed reliable? In classic survey response methodology, a response rate of 60%, though preferably 80%, is seen as acceptable for analysis and generalization purposes (Dillman, 2000). With the 14% response rate at the AAM study, one could repeat the exact study next year and have an entirely different 14% respond with entirely different answers. Whether the implementation of mobile projects had gone up, down or remained constant, it would be impossible to say reliably. For the Pocket Proof study, we cannot calculate a response rate, as we don't know how many individuals saw the survey request. We can not say with certainty what percentage of museums are currently conducting a mobile project, planning a mobile project, whether mobile projects are more common in certain types or sizes of institutions. While the concept that mobile interpretation is more prevalent in large art museums has significant face validity, due to the combination sampling strategy and response rate we have no consistent, reliable numbers.

While the results within these studies may not be generalizable, they still provide some valuable insights. One of the most useful aspects of both studies is the documentation of barriers faced by museums in implementing mobile projects. The percentages aren't necessarily relevant here, but the rank order is. These lists provide institutions a list of most prevalent obstacles.

Setting aside the idea that some set of museums underrepresented within the surveys might face significantly different obstacles towards implementing their mobile project, the barriers faced by those that are already engaged in such projects, and by those contemplating such projects, is illuminating. As the Pocket Proof/Learning Times study notes, for institutions already using mobile interpretation, encouraging visitors to use the mobile interpretation was the largest challenge. Yet for others—vendors and researchers, as well as those considering projects—attracting new visitors via mobile was a primary goal. This disconnect represents a great opportunity for future research. Despite the numbers of institutions exploring mobile and the availability of phones, usage rates remain below 10% for permanent galleries (Proctor, 2010) For those contemplating employing a mobile interpretation strategy in the Pocket Proof study, the lack of visitor interest was at the bottom of the list of obstacles, whereas for those currently offering mobile interpretation, this issue was more of a concern. That many visitors simply find the concept of using their phone unappealing in this context (Haley Goldman 2007) is a finding that must be further explored. [1]

## Mobile Research Opportunities

The groundwork laid by these studies promotes reflection on what sort of studies are still needed. Both of the studies here had the foresight to ask their participants what type of research is still needed within the industry. Building on those concepts from a researcher's perspective, I propose three avenues of future potential research.

### 1. Visitor-based research

I know from personal experience that evaluation and research on visitor use of mobiles in museums is extraordinarily difficult. The small sample size makes it difficult to get a full picture of usage, the interpretation strategy creates difficulty in finding those who are using their mobiles; the list of challenges goes on and on. The research on mobiles in museums is overwhelming institutionally-oriented, as is the motivation for many of the museum implementation projects. Some institutions wish to avoid the overheads of providing audio tour devices to visitors, others wish to try to engage their public in more technologically current ways. The research described here focuses on the museum, their project readiness, their motivations and their obstacles. The visitors have a significantly different perspective, including their own motivations and barriers to use. The gap in visitor use can be examined through a number of lenses: how individuals adopt uses of technology, how visitors perceive their phones, and visitor motivations and desired outcomes for their museum visit.

Information-seeking is one of only many potential uses of an individual's phone (compared to social utility, accessibility, status, etc.), and is not by any means the most common use (Wei and Lo, 2006). Thus visitors' perception of their phones does not immediately indicate the phone's usefulness as an interpretative device. Whether visitors are likely to use their phones for interpretation depends on their goals for their museum visit.

Similarly, while learning is a key element in many visitors' articulations of why they choose to visit a particular museum on a particular day (Kelly, 2000), it is not the only motivation for a visit. Visitors come for destination visits (been there, done that), to spend meaningful time with family or guests, and for many other reasons. (Falk, Moussouri & Coulson, 1998). If the use of their phone for interpretation does not immediately further their goals for the visit, visitors will not make use of the opportunity. In-depth *qualitative* research exploring the relationship between visitor motivations, expectations of the mobile product, and resultant museum experience would help developers create better visitor personas. These personas, based on how visitors have adopted the technology, as well as their motivations and social groupings, would create better products.

## 2. Case Studies

As tempting (and useful) as it is to cast a wide net to look at mobile adoption across the field, analysis at a micro-level, both institutionally and from a visitor perspective, would be extremely illuminating. The AAM study notes case studies (and visitor research) at the top of the list for future studies. The generation of case studies would be an excellent complement to the analysis of barriers that each project might face. The AAM study notes that there are widely divergent goals for these projects, from increasing visitor engagement to marketing to bringing collections to a broader audience. Separate case studies by project goal would allow institutions to focus on the strategies most appropriate for their goals. The call for case studies is not new, and yet the number of viable case studies compared to the estimated number of projects occurring is very small. Pulling together a case study is quite difficult when deeply embedded within a project, and even more difficult to do in a way that is comparable with case studies from other institutions. A single collaborative project or effort to research and generate case studies from multiple institutions would provide the most comparison points for other institutions.

## 3. Stratified and/or Longitudinal Field-Wide Studies

For future studies designed to look at the landscape of mobiles in museums from an institutional perspective, perhaps rather than casting a field-wide net, it would be more profitable to cast a series of smaller, more fine-grained nets over time. Given that the museum field is so vast (the AAM study had thousands of responses, and yet that only results in a 14% response rate), future field-wide approaches should invest in stratified or longitudinal studies. A study of a stratified sample of museums would allow closer examination of factors that may influence development of mobile products, such as number of visitors or content focus of the institution. Longitudinal studies would also allow a more reasonable sample size, but could chart the change in mobile developments over time. Do institutions face the same barriers they did a year ago? For those just contemplating a mobile product, were they able to create one or are they still in contemplation?

In conclusion, there is much to be done, both on the product development portion of the mobiles in the museum field, and in the research components. These studies provide any important first step. The next steps should perhaps be narrower, but deeper and much more deeply tied to the visitor.

**Note**

1. Studies at the Whitney (EMCarts, 2008) and other places (Haley Goldman, 2007) provide the beginnings of insight into visitor non-use of mobile interpretation, suggesting that while logistical issues such as minutes, lack of connection and battery time, account for a portion of visitors disinterest in utilizing the mobile capabilities, a significant portion of the reasons for visitor non-use remain unexplored.

| 9 | JANE BURTON |
| **Playful Apps** | |

THERE are hundreds of thousands of apps for smartphone consumers to choose from, and most of them are games. Games make up 70 to 80 percent of all apps downloaded. The latest reports say that 26 million people spend at least 25 minutes every day playing games on their phones [Flurry Analytics, Feb 2011]. The incredible amount of innovation in smartphone mobile gaming is showing us how to create content that people really want to spend time with. The question museums and galleries need to answer is whether there is room in this marketplace for "serious games," games that offer more than just pure entertainment.

The potential to bring significant ideas to life within the framework of game-play is something that has been brilliantly expressed in the work of Jane McGonigal, author of "Reality is Broken," and by UK television's Channel 4 Education team, who have put gaming at the heart of their content offering to young audiences. But little of the research and innovation around "serious games" has focused on the rapidly growing area of apps.

One of the most successful games produced by a museum is Launchball, from The Science Museum, London, which was first developed in 2007 to be played on kiosks in gallery and online. Designed for children ages of 8 to 14, the game requires you to guide a ball through a series of fiendish challenges, using fans, magnets and Tesla Coils to help you as you learn basic scientific principles along the way. The online version proved so popular, gaining 5.3 million players, that in 2009 the Science Museum re-launched it as a paid-for app. So far the app has been downloaded 7,842 times, enough to pay for its development and return a modest profit, says the museum. This is an achievement, given that the same game can still be played for free online. Nonetheless, the disparity in the figures is a striking reminder of the reach of the web compared to any given app store, and of the power of free content.

Another great "playful" offering, though not strictly a game, is the MEanderthal app from the Smithsonian's National Museum of Natural History, which transforms your photo into the face of an early human. You upload a photo of your face, then choose which human species you'd like to become as you morph back in time. There is a serious point behind the fun: "We think it's really important for people to make emotional connections to our ancestors," commented Briana Pobiner, a paleoanthropologist at the Smithsonian. "It's an important way to break down that barrier between things we think are so different or so 'other.'"

At Tate, we're interested in finding out if app gaming mechanisms can be applied to an art context. We have produced one game so far, and have two more in development. Launched in 2010, Tate Trumps is a digital card game you play with the art on display at Tate Modern. Visitors can download the game for free to an iPod Touch or iPhone, roam the galleries, choose seven high-scoring artworks, and then play a fast-paced and strategic game of Trumps. There are three different modes (Battle, Mood, or Collector) and you can play on your own or with your friends or family. In Battle mode, you need to ask yourself the question, "If this artwork came to life, how good would it be in a fight?" In Mood mode, you're looking for artworks you think are menacing, exhilarating or absurd. Or, if you wish you had a gallery of your own, try Collector mode, and find pictures that are famous, recently produced or practical to house.

Tate Trumps is unashamedly light-hearted, but at its core promotes the acts of discovery and looking—key to any art experience—whilst encouraging

Only one player can collect each artwork, so try to grab good ones early.

Back    Continue

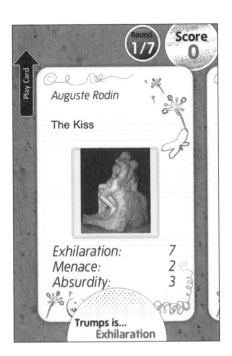

Round 1/7    Score 0

Play Card

Auguste Rodin

The Kiss

Exhilaration:    7
Menace:          2
Absurdity:       3

Trumps is...
Exhilaration

people to form their own opinions. Being a multi-player game, it also acknowledges the fact that gallery-going, for many, is a social activity, shared amongst friends.

Tate Trumps was deliberately designed to be played only at Tate Modern in order to encourage a direct encounter with artworks. But of course app stores are global marketplaces, and we hadn't reckoned on the frustration that not being able to play would engender in the majority of people who wouldn't be coming to the gallery in the near future.

For our next game, currently in development, the brief was to come up with something that could be played anywhere, without visiting the museum, but with bonus content for those who can make it to Tate Modern. Shake your phone and the "Magic Tate Ball" will curate a piece of artwork that relates specifically to that unique moment in time. Pass it round the pub or check it on the train to find out which artworks fit the DNA of your daily life. In auto-mode, the application will use the iPhone's sensors (microphone and GPS), along with other feeds like weather and time, to deduce the most appropriate artwork for the given criteria. In manual mode, the user can ask the Magic Tate Ball to generate ideas on themes: Inspire me; Shock me; Give me a Talking Point.

The third game we are developing pushes further into pure gaming territory. The challenge we've set ourselves is to take a simple, addictive form of gameplay along the lines of Doodle Jump and bring art into the mix, imparting meaningful information without getting in the way of the action.

The jury is still out on how successful games like these will be in terms of introducing new audiences to Tate's Collection, and we will be evaluating them later this year. But in the meantime, here are a few things we've learnt along the way.

### Know Your Audience.

It's easy to assume that mobile gamers are teenage kids. Wrong. Forget the acne generation, unless you're talking about console platforms like Xbox or the PS3. The typical gamer downloading games through app stores (and really, we're still talking about iTunes, though Android is beginning to build a market share) is female, between 18 and 49, and well educated. A recent report published by Flurry, a San Francisco-based smartphone analytics firm, said: "Studying the U.S. mobile social gamer, we note that she earns over 50 percent more than the average American, is more than twice as likely to have earned a college bachelor's degree, and is more likely to be white or Asian." The number of men playing isn't far behind, though: 47 percent of app-based gamers are male, compared to 53 percent female. In fact, the profile for these gamers is strikingly similar to the profile of many museum visitors, which suggests that app gamers may very well be open to cultural content delivered in this form.

However, this demographic will widen as Smartphones become more affordable and therefore more common over the next two years. If you want to use mobile games to reach a teen audience, start planning, but maybe not developing just yet, and look beyond the iPhone platform.

*Make it Free.*

With hundreds of thousands of apps available for mobile consumers to choose from, it's a tough market, and most publishers are moving towards free apps. Some are supported by sponsorship, or possibly, if you've got a really hot property, by "freemium," whereby you get the basic app free, but people pay for the fancier version. Anyone who has played "Angry Birds" will recognize this model. But only the most optimistic developer from the cultural sector would imagine they are going to make much money from a game.

*Think about Discoverability.*

Submitting your game into an app store is a bit like dropping thousands of dollars down a well. There's an initial splash, which dwindles to a ripple, then silence. The iTunes app store is a busy place, and it's hard to get noticed amongst the crowd. Getting noticed is known in the digital world as "discovery," and there are myriad social networks and recommendation sites springing up that aim to make app discovery easier for consumers. Some of the sites are listed here. But the sands are ever shifting, and there are no sure-fire solutions. Being early to market in one of the less crowded app stores looking to rival iTunes is beginning to look like a smart move.

10

**So Many Devices,
So Many Options:
An Introduction to
Cross-Platform
Thinking**

ALLEGRA BURNETTE

TRADITIONALLY we are trained to think that form follows function: first you decide the content and purpose of what you are designing, building or creating, and then you shape the form around that. But in many ways, mobile projects are the reverse—the function of your app is often led by the form factor of the device itself. A tablet, for example, offers a different kind of experience than a smartphone, and that difference can shape the app you develop.

Smartphones are inherently portable, enabling us to slip a vast array of content, activities, and resources into our pockets or bags and take them wherever we go, pulling them out when we need to look up something or find out where we are. They are therefore ideally suited for things like audio tours in the museum, finding out the location of a museum and the next event or film showing, and looking up an art term while standing in front of a work of art. Going beyond the one-way communication of a traditional audio tour, mobile phones enable two-way communication between the museum and its visitors, as well as visitors with other visitors.

Tablets, while also portable, are (at least currently) typically much less about a literal on-the-move experience—even Apple's ads when the iPad came out last year showed someone with the device propped against his knees, leisurely perusing an app or browsing the Web. People use tablets while traveling on planes, trains, and in automobiles. They use them to read magazines, play games, and watch a movie. A tablet can be viewed by several people at once more easily than a phone can. But at the same time, the tablet format creates a more intimate personal space between user and device than a laptop or desktop computer does: we are interacting directly with the screen rather than through a separate keyboard or mouse, and we are often holding it rather than facing it.

At The Museum of Modern Art we prioritized the iPhone as our first platform for mobile development because iPhone users were the largest mobile audience of our website. (While Android was second largest at the time we set our priorities, the iPad overtook the Android audience shortly after it first came out in the spring of 2010, and remains the second largest at the time of writing.) In order to make the best use of our resources and to streamline development and ongoing maintenance, we developed an app that was a hybrid of a native app and a mobile site. Creating this hybrid app rather than just a mobile site gave us access to broader distribution through the Apple App Store. This also gave us the opportunity to create features like MoMA Snaps, a branded postcard activity, which would not have been possible through a browser version alone. But at the same time, it allowed us to develop a structural base that could be adapted for an Android app and the mobile version of our website, MoMA.org.

The MoMA iPhone app that we developed was meant as both an in-museum and an offsite experience. Like many museums currently developing mobile apps, we wanted to include our audio tour content. But we were not intending to replace or supplement the current in-museum audio devices with iPods loaded with the app (due to the quantity needed, as well as distribution, security, and maintenance issues). Instead, our intention was to offer the app for people who wanted to access the content through their own devices when in the museum. We included the entire calendar of events and exhibitions and access to all of the online collection with the intention that people would also use the app to plan a visit or learn about works of art beyond the walls of the museum.

While the iPhone app (and the later Android and mobile versions) was a more general view of MoMA and its collection, the iPad app we developed for the *Abstract Expressionist New York* (AB EX NY) exhibition was an exploration in creating an experience specifically for a tablet device around a single exhibition. We chose not to do a tablet version of the mobile phone app immediately because we felt the form factor necessitated a different approach to the content. This initial tablet project gave us a chance to explore focused ideas on how we could present works in our collection, which in turn might later inform broader, tablet-based projects that we may develop in the future.

Several ideas we explored in the tablet format would have been less effective or not possible in a phone-based app including, for example, a split screen layout, which allows textual information to appear adjacent to a work of art. It is very difficult to combine text and art onscreen in a meaningful way on a phone—your focus is either on one or the other (which is in part why video and audio are particularly effective on a smartphone).

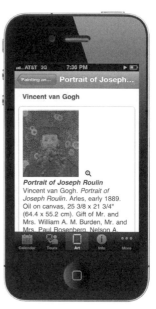

The home screen of the iPad app was a scrollable view of the works that were featured in the program, with the images shown loosely in scale with each other. This selection, combined with the "gallery" browse views, creates a different experience than the typical, more list-based phone app.

Mobile Apps for Museums

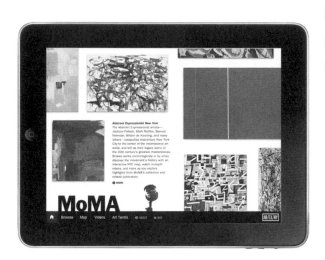

While these are certainly not all of the different layout considerations between a tablet and a phone app, they do hint at the larger issue at hand: How do we create compelling experiences for the different device form factors with the limited resources available to museums and in the rapidly changing face of technology? The sand is shifting so much right now that there is not currently a clear answer, but being strategic and thoughtful about how you approach the various platforms (tablet versus phone) and formats (app versus Web), while staying true to the content and your own capabilities (or those of a trusted consultant), is at least a start.

While the AB EX NY iPad app was intended to promote the exhibition, its related publication, and MoMA's collection, we very specifically intended it to be an experience that took place outside of the exhibition, whether that meant people used it before or after a visit, or even if they never came to the museum at all. We even used images inspired by the Apple campaign of someone using the app in a non-museum space to reinforce that idea.

Even though the app includes the content from the audio tour, it really didn't occur to us that people might try to use it within the museum as a mobile app, until we read this in a review:

> You may find the experience of lugging an iPad around the exhibit distracting; I certainly did at times, for no other reason than all the attention it attracted. But if you think about this as a piece of software, free to be downloaded onto any iPad anywhere with an Internet connection, then it dawns on you: a kid in Idaho, two time zones and two thousand from the MoMA, can experience this content as easily as a youngster from the Bronx.[1]

This was a valuable lesson: no matter how we design these apps and no matter how carefully we tailor them to a particular platform, the known unknown is how and where people are going to use their mobile devices. Anecdotally, we have noticed that in the museum, people are using both phones and tablets, with phones making more of an appearance in the galleries and tablets used more in the interstitial spaces. But if we offer the same program on both phones and tablets, would visitors switch between devices based on where they are, or are they more device-consistent within the space of the museum? At this point, only more observation and testing will reveal the answer to that.

If we look at the number of mobile phone apps versus tablet apps in the Apple App Store and Android Market, we see far more of the former than the latter. While this is in part due to the fact that tablets are newer to the market and comprise a smaller share of the mobile device landscape in general, it may also be due to the different experience of using a tablet and the different requirements, including interface design, needed to develop those experiences. And while "function follows form" may be in large part the way we've started developing mobile museum apps, as tablets start to come out in varying sizes and phones screens get larger, the differences between a phone versus a tablet experience is likely to become blurred. Add to that the various ways that people use mobile devices, and there are more overlaps between smartphones and tablets. Careful planning and evolving development solutions should help clear a path through the morass as we move past our mobile beginnings to a multiplatform future.

**Note**
1. http://www.pcmag.com/article2/0,2817,2373284,00.asp

| 11 | Koven Smith |
| --- | --- |
| **Mobile Experience Design: What's Your Roll-Out Strategy?** | |

RAPID advancements in smartphone technology of the last few years have changed the nature of mobile experiences in museums utterly. Where tour-based audio guides were once the only type of mobile experience available to museum visitors, we are currently witnessing an explosion in the types of experiences from which visitors might choose. Augmented reality games, crowd-sourced content creation, or even experiences not designed to occur inside the museum at all are just a few of the new ways that museums are beginning to explore to enhance either a physical or virtual visit.

These new opportunities mean that museums must now take a more nuanced approach to how their mobile experiences are introduced to the public. In the past, the expense of providing mobile experiences to visitors (typically via audio) meant that those experiences needed to appeal to the broadest possible audience in order to make them cost-effective. This broad appeal was reflected in the brute-force marketing strategies employed by museums to encourage uptake: handing out mobile guides to visitors, advertising the guides with large signs at the entrance, and often providing mobile guides as a premium benefit of membership.

However, development of mobile applications and mobile web sites as replacement for dedicated devices as the primary means of delivering mobile experiences—and the concomitant reduction in production costs—means that mobile experiences in museums no longer need to be designed for a museum's entire audience in order to be cost-effective. Many of these new types of mobile experiences are often aimed at a particular niche audience, whether scholars, gamers, children, or social butterflies. Each of these niche audiences requires its own type of solicitation, both via the design of the mobile application itself as well as the marketing campaign used to introduce it. Museums must therefore design the strategy by which a mobile experience is "rolled out" to the public as carefully as it designs the mobile experience itself. The goal of a successful mobile roll-out strategy should not be to reach *more* users, but rather to reach more of the *right* users.

Reaching the right users involves reflecting the needs of a given target group in the design of the application, but also in the ways the target group is approached to participate. Mobile applications designed for a small subset of a museum's public shouldn't be marketed to every single person who walks in the door, any more than an application designed for use by the "average" visitor should be marketed exclusively to the gaming community. A museum's roll-out and marketing strategy should act as a signal to visitors indicating what type of experience they should expect; visitors should then be able to better self-select the kinds of experiences that are right for them. What follows is a discussion of three typical roll-out strategies for mobile experiences in museums, with a discussion of how the target audience, application design, and marketing strategy affect one another. These strategies should serve as solid starting points for any museum contemplating how to introduce its visitors to a new mobile experience.

## Scenario 1: Broad Appeal

### Target Audience

In a "broad appeal" scenario, the museum is marketing its mobile experience to every single visitor who enters the building. A mass-market roll-out scenario is designed to reach the largest number of potential users, typically from a wide array of demographic backgrounds. In this scenario, the visit to the museum drives use of the mobile application; the average user has probably not arrived at the museum already aware that a mobile experience is available to him or her, necessitating a wider-reaching information/marketing campaign. The typical result of this kind of campaign is a relatively passive type of engagement from a large number of users.

## Design

If a mobile application is to be marketed to the masses in this way, it must be truly usable by those masses. The application should be highly fault-tolerant and forgiving of mistakes on the part of the visitor, not dependent on the user's familiarity with other technologies or services in order to participate in the experience ("sign in with your Twitter account" would be a poor way to kick off the experience, for example), and not contingent on the user's familiarity with specialized language or jargon. Because the user engagement in this type of scenario is likely to be low, the threshold to content consumption should also be low.

## Strategy

Because this type of experience is designed for most (if not all) visitors to the museum, the roll-out of the experience should reflect this more "populist" nature. It should be impossible for any visitor to exit the building without knowing that there is a mobile experience available to him or her. The most straightforward way to encourage adoption is simply to rent or loan the visitor a device with the application pre-loaded on it. Prominent signage and other kinds of promotional materials (e.g., bookmarks reminding visitors to rent a tour, or offering a discount) at the entrances and in dwell spaces will also help to saturate the physical space with the awareness that the mobile experience is available.

In instances where giving a device to the visitor may not be possible, the museum must make the application available for download to users' own devices. Making an application available in this way is not as straightforward as it might seem. First, signage must be available—at every location that the visitor might use the application—directing the visitor how to download the application to his or her personal device. Many users may not download the application at the front door, so it is important to have additional signs prompting download throughout the building, in locations where content is available. The nature of this signage should also reflect the nature of the application design. If the application is primarily aimed at researchers, scholars, or students, for instance, a broad appeal campaign may not be appropriate: the museum cannot create a broadly targeted information campaign for an application that will be difficult for all but a small minority of its visitors to use. A large sign saying "Download our mobile app!" is a not-so-subtle message to the user that the app will be easy to use, and will not require much from the visitor.

## Scenario 2: Stealth

### Target Audience

A "stealth" roll-out means that the museum has decided to market its mobile experience to a niche group without an overt information campaign. The expectation is that this group will be a subset of the museum's total visitor profile, but that this smaller group will be far more actively engaged with the mobile experience than the "average" visitor. With a stealth campaign, users of the mobile application should already be aware of the application before a visit is made, if the application itself did not in fact prompt a visit.

### Design

A "stealth" campaign is an appropriate means of marketing when discovery, exploration, and mystery are primary components of the application design. While the application shouldn't necessarily be difficult to use, the act of figuring out how the application works should be a key part of its appeal. Because a stealth campaign is targeted at a smaller audience, the application should have an appeal tailored to the audience being targeted. An application designed to be a broad introduction to the museum's collection, for instance, wouldn't necessarily benefit from being rolled out in this manner.

*Strategy*

In a stealth campaign, the marketing of the mobile application is itself part of the total experience. The primary goal is a highly engaged user community, so the roll-out campaign should be designed to promote a high level of curiosity at the outset. There are a number of strategies a museum might employ in pursuit of this goal. A straightforward strategy might be to identify potential "influencers" in the museum's community, and give those influencers a personal introduction to the application, with the expectation that these users will provoke others to use the application as well.

Another possibility would be to take a cue from alternate reality campaigns, and attempt to promote a sense of mystery around the application. In this scenario, the museum might use signage, but in a more oblique way than in a broad appeal campaign. The museum might embed "clues" that would prompt a visitor to download the application to his or her phone or to take additional actions. Clues could even be embedded in materials designed to be consumed *outside* of the museum, such as print materials or the museum's web site. An effective stealth campaign should guarantee user interest and engagement long before the application itself is actually downloaded.

## Scenario 3: Third-Party

The recent explosion in the number of museums making collections content available via APIs ("application programming interface") has created a third viable scenario for museums: an application designed by a third party, completely outside the museum's purview and control. Strategizing for both design and marketing of a mobile experience developed in this way is challenging, but not impossible. A museum in fact has the ability to influence both the design and introduction of a mobile experience, even when developed largely without that museum's input. The target audience of this type of experience is variable, depending on the museum's goals.

## Design

Again, in this scenario the museum is looking for means by which it can influence the development of a mobile experience more than overtly control that development. Here, a museum might look at multiple types of content to make publicly available. A museum making its information available via an API might create a separate "mobile-ready" API that prioritizes the types of information the museum would like to see in a mobile device. Delivering data in this way helps to ensure that the mobile application developed by an outside developer will still focus on the kind of information that is important to the museum. A museum might also publish a list of locations within the building, with particular content attached to each, or a database of artists within the collection, or a geotagged list of artist birth/work locations. Museums might make any number of content types available that would encourage developers to create applications that travel outside of the normal "tour" format.

## Strategy

A third-party mobile experience represents a unique challenge for museums from a roll-out and marketing standpoint. Because the museum may not know that an application is being developed until it is already publicly available, it is difficult to schedule its roll-out into a broader marketing strategy or schedule. In this scenario, what the museum should be looking to do is to provide incentives to potential mobile developers to work within the museum's ideal framework. There are a number of ways a museum might do this. It might simply indicate a willingness to promote an application in its galleries or on its website if the application is developed according to certain standards. It might also be willing to provide physical infrastructure for certain types of applications (AR or gaming applications, for example). While doing this, it is critical that the museum keep in mind how to create brand differentiation between its own "official" applications and those developed by the community, inserting appropriate language into "terms of service" agreements and the like.

## Conclusion

A diversity of mobile experience types demands a parallel diversity of marketing approaches. It is clear that museums need to begin making far more deliberate choices about how their mobile experiences are rolled out to the public. Making the right choice will help to ensure that the right visitors are paired with the right types of experiences. Whether the museum wishes to reach its visitors outside the building, via a game-style approach, or inside with a mobile tour, the roll-out strategy should be carefully considered at each and every stage of development.

| 12 |
| --- |
| Enhancing |
| Group Tours |
| with the iPad: |
| A Case Study |

ANN ISAACSON, SHEILA McGUIRE,
SCOTT SAYRE, AND KRIS WETTERLUND

## Introduction

The Minneapolis Institute of Arts is an encyclopedic art museum with a permanent collection of around 80,000 objects. Currently, 385 volunteer tour guides
in the museum's Department of Museum Guide Programs interpret these collections for over 140,000 visitors annually, helping fulfill the museum's mission to
make the collection accessible to the community.

Digital media has been a long-time friend of the early-adopting Minneapolis
Institute of Arts and its visitors. It started with the installation of gallery-based
kiosks in the early 90s, then an early handheld tour in 1997 (using Apple's
Newton), cell phone tours in 2001, and a tour app in 2010 (Sayre, 1993, 2005,
2007). While all of these technologies were designed with the visitor in mind,
the museum's innovative tour guides have found ways over the years of incorporating components of these programs into their tours. For example, tour guides
frequently gather their groups around gallery kiosks to show videos.

Interest in media integration was rekindled during a 2008 Symposium
session on integrating digital media into tours. Tour guides passed around an
iPhone playing a video of artist Dale Chihuly creating a chandelier identical
to one in the lobby at the Minneapolis Institute of Arts. The tour guides were
"ignited," remembers one participant, about the possibilities of presenting
dynamic media literally in the palm of their hands. Contrary to popular trends,

the tour guides' excitement was not about handheld technology itself, but the potential of presenting portable media on tours to enhance visitors' understanding of works of art.

Following up on the potential of this experience, Wetterlund and Sayre began to explore the possibility of mobile social computing, in particular the human-guide-mediated experience (Sayre, Wetterlund, 2008). After reading their article on social computing, the MIA contacted Sayre and Wetterlund about the integration of portable media in their tour guide program. The museum understood what visitor research has shown: the majority of visitors come to the museum in groups (Draper, 1984) (Sachatello-Sawyer et al., 2002). The announcement and release of Apple's iPad was timely in the delivery of a user-friendly solution.

------

## Research

With the right computer in place, a research project was developed to explore the feasibility of enhancing group tours with a range of rich media delivered via a single iPad, used as a group presentation device. The project defined three primary areas to explore.

### Response

- Visitor response
- Tour Guide response
- Museum educator response
- Museum-wide response

### Obstacles

- Political and psychological obstacles
- Physical obstacles
- Technical obstacles

## Logistics

- Training needs
- Material preparation and organization
- Most effective materials
- Hardware management

## Instruments

The research team developed three instrments to assist in the collection of data:

- Observer checklist and comment document
- Pre-training survey
- Post-training survey

## Project Phases

The project research was divided into five phases:

- Media and App Selection
- Trial Tours and Observation
- Resulting Strategies
- Tour Guide Training
- Survey Results

## 1. Media and App Selection

Early on in the process, the educators sent an e-mail inquiry to all of the museum's volunteers. This yielded an abundance of creative ideas about enhancing tours with multimedia, and a healthy dose of concern that the technology might distract from the artworks. The most popular responses included a desire to have world maps, music, pictures and diagrams, and videos of art processes, dance, artists, etc. The team collected a range of media set up on a PC to serve as a hub device from which all of the iPads could be synchronized.

The team also investigated apps for organizing, navigating and presenting the various assets on the iPad itself. In the end there was no one app that could manage all forms of media. The team settled on managing audio and video as playlists within the iPod app, images in the Photos app, and the Safari and Google Maps apps for bookmarked locations. Important aspects of an iPad file management app for tour guides include:

- Integration and linear (swiping) presentation of all forms of media, in both landscape and portrait mode.

- Integration with iPads' onboard media assets managed on the computer to which the iPads are synced.

- Ability to organize assets on the iPad based on tour type, subject or tour guide.

- Light table and search capabilities.

- Ability to create multiple shortcut "aliases" to the same onboard media asset in two more locations, so that the assets themselves do not need to be duplicated.

- Ability to add metadata to media so it can be easily queried.

## 2. Trial Tours and Observation

Several tour guides volunteered to try the iPad with tour groups, accompanied by a staff evaluator who used the observation checklist to record events on 20 tours. The tool was designed to gauge the effectiveness of the iPad based on the engagement of the visitors. The observation tool assessed:

- the portability and presentation strategies for the iPad;

- whether the tour guide was able to successfully navigate the different types of content on the iPad;

- whether the content selected for presentation was integrated appropriately into the topic of the tour;

•whether the quality of the content (audio, video, etc.) was adequate for the tour group;

•whether all visitors on the tour were fully engaged with the iPad content when it was presented.

Portability of the iPad was not an issue. In some cases, the iPad was a relief, as it took the place of larger, bulkier props. In all cases, the tour guides successfully navigated the content on the iPad, easily locating different types of pre-loaded content, easily adjusting the image brightness and volume. The screen size, brightness, audio level and quality, and video and image quality were rated excellent to good.

Most important, all of the museum visitors were engaged during the iPad portion of the tour. All visitors indicated understanding how the iPad content related to the tour content, and all thought it added to their understanding of the works of art. Visitors responded to short videos illustrating artistic processes or techniques with an audible "ah ha!"

### 3. Resulting Strategies

The initial rounds of in-gallery testing yielded a number of useful strategies. Like any gallery prop, digital media should be used judiciously to avoid making it the focal point of the tour. For example, while facilitating a discussion on a Lakota beaded dress, a video demonstration of beading technique could be introduced: "Now that you've had a chance to see the fine detail on this Lakota dress, let me show you a video demonstrating how those perfect rows of beads are sewn onto the hide." Explaining any kind of prop, traditional or multimedia, by cluing visitors on what they are about to see and explaining what they are likely to get out of the prop is good educational practice.

Holding the iPad so that the screen faces away from the visitors while searching for information relieves them of the burden of having to watch while the tour guide taps around the screen. A thoughtful, open-ended question can momentarily turn the focus away from the mechanics of locating information on the iPad.

Displaying media on the iPad about shoulder to chest high seems most effective. When not in use, darkening the screen avoids diverting attention away from the tour presentation. Video or audio is best kept short, between 30 to 60 seconds. While the iPad's speakers are remarkably good, hearing audio or video in a crowded gallery can be problematic, especially for larger groups. Tour groups at the MIA typically range from 10 to 15 people. MIA tour guides solved this problem with larger groups by turning the sound off and narrating the video themselves.

### 4. Tour Guide Training

To take advantage of Apple expertise and hardware, a training for MIA tour guides was offered at a local Apple store to introduce the iPad as a hardware device, and help the tour guides understand how to navigate the interface. The Apple business team was excited to learn about the MIA project, and over 50 tour guides attended these sessions.

The next training sessions were held at the MIA. Nearly 100 tour guides and Apple Store business team employees attended these sessions, where the goal was to make it clear how to access content organized on the iPad, and to model the strategies discovered in the trial tours for presenting multimedia on the iPad. In addition, tour guides went over the procedure for scheduling and checking

out the iPads from the office, as well as the docking procedure when the iPads are not being used.

### 5. Survey Results

An online pre-training survey was distributed to all tour guides interested in iPad tour training. The survey collected information about the tour guides' experience with the museum, technology, and their preconceptions about using the iPad in public tours.

Key findings showed (N = 97)

The majority of the respondents had over 10 years experience as guides.

• 47% had previously integrated gallery media content into their tours.

• 74% considered themselves to be experienced computer users (69% PC).

• 30% owned smartphones.

• 62% had never used an iPad.

• 20% had attended the previous iPad device training, and 80% felt it was helpful.

•90% who attended the iPad device training were interested in using the iPad in tours, with the remaining 10% unsure.

Preconceived advantages of using the iPad in tours:

Images/Zooming: 47%, Looking up information: 18%, Video: 16%, Maps: 13%, Music/Sound: 8%

Preconceived concerns related to using the iPad in tours:

Overuse/distracting: 37%, Nervous/Performance: 37%, Image too small for large groups: 16%, Gallery noise: 3%

An online survey was conducted approximately 30 days after the iPad tour training. Post training survey participants (N = 49) were correlated against pre-survey participants according to their email addresses, resulting in a total of 38 (10% of the total volunteer group) who participated in both surveys.

Key findings showed (N = 38)

•66% had used an iPad since training.

•8% had purchased an iPad since training.

•24% planned to buy an iPad.

•21% (8) had given one or more tours using the iPad.

•63% of those who attended the tour training were planning on using the iPad in their tours, with another 34% unsure.

•91% of the trainees who attended both the iPad device and iPad tour training were planning on using the iPad in their tours.

•26% felt they could benefit from more training, with 42% unsure.

Suggestions included mentoring, practice sessions, tour guide group presentations.

•39% felt they could benefit from more support, with 34% unsure.

Suggestions included loading and locating content, work process and identifying gallery objects with additional content.

Preconceived concerns related to using the iPad in tours:

Overuse/distracting: 21%, Image too small for large groups: 21%, Nervous/performance:16%, Taking too much time: 8%, Gallery noise: 5%, Fragility/breaking: 5%

The key conclusion from the surveys was the effectiveness of combining the iPad device training with the iPad tour training. Tour guides who participated in both training sessions demonstrated a much greater degree of confidence in using the device and integrating it in their tours.

---

## Discoveries

The iPad tour project has encouraged staff throughout to think of new content to integrate into the program. MIA photo services staff suggested additional "hidden" details to include, and MIA curators brainstormed an array of exciting possibilities. They also look forward to using the iPad to share whole print series, complete ledger books, and other resources that cannot be viewed in the galleries.

The PC hub computer has become a community repository for content as staff and tour guides are able to contribute music, videos, and photos. Everyone using the iPads has access to the materials contributed by their colleagues. Also, the volunteers who are excited about the media that can be integrated into tours with the iPad are the most powerful advocates for getting their peers engaged in the process. The media content on the iPads has inspired tour guides to include objects on their tours that that they would not have considered before.

Another welcome outgrowth of this initiative is that visitors become active participants in the creation of the digital stories being told during the tour. For example, on a recent tour a visitor made a connection between a terracotta portrait head from the Ife Kingdom and the ancient Egyptian queen Nefertiti, but was unable to relate the two chronologically. A search on the iPad put Nefertiti's dates and images at the tour guide's fingertips, contributing a whole other level of user-inspired content to the discussion and validating the visitor's contribution.

In the early planning for this project, the team was most excited about the possibilities for video on the iPad. In practice, however, pictures on the iPad prove the most powerful. Visitors are delighted when a small object encased in Plexiglas in a dark gallery appears on the iPad, and the tour guide zooms in to show the details.

Photographs of things that are not possible to see in the galleries are also riveting for visitors. Detailed photos of the engine of a car when the hood is closed in the gallery, an embroidered chest with all of its drawers open, or the underside of a vessel all have visitors studying both the object on view and the iPad intently.

---

## Challenges

For a decade and a half, tour guides have expressed concerns that the museum might want to replace them with technology. But in the end, especially with the integration of technology into their tours, tour guides are assured that people want human interaction. Far more visitors interact with tour guides annually than take advantage of audio tours. Peter Samis has it right when he describes humans as the "ultimate interactive device," context sensitive, and responsive to questions in real time. (2007)

Some of the challenges the project team encountered are more obvious: iPads are expensive, many volunteers are inexperienced with or fearful of the technology, and managing a lot of media files can be cumbersome and intimidating, especially in front of a group. The greatest technical challenge on tours has been the volume of the iPad when the galleries are full or the group is very large. A case with built-in speakers would be an ideal solution.

## The Future

Challenges aside, the enthusiastic response from all of the stakeholders has set a course for continuation and expansion of the iPad program at the MIA. The MIA is currently using Apple products in its tour guide programming, but other tablet computers are coming to market that will likely offer similar potential and perhaps even more possibilities

New hardware features like the iPad2's cameras will make it possible to experiment with bringing voices outside of the museum into tours in real time. In the future, tour guides might engage visitors in conversations with artists in their studios, or with children in museums in other parts of the country. The cameras can also be used to capture QR codes from museum labels to help tour guides quickly access related content.

New technology like Apple's Airplay allows tour guides to send media stored on an iPhone or iPad wirelessly to a larger dedicated or multipurpose monitor or projector connected to AppleTV for group presentation. New apps for better managing, organizing and presenting all forms of media content on the iPad continue to be released and assessed. Recent prospects include Best Album, which provides tools for organizing, cataloging, searching images, video and audio within personalized albums.

And finally, once existing media has been mined for its potential, the logical next step for museums is to undertake the production of new media assets specifically for use in iPad-enhanced tours.

Interpretive techniques are expanded when considering the potential of presenting multimedia on tours. Hopefully other museums will embrace these opportunities and help foster a new era in museum tour experiences.

## References

Draper, L. (1984). "Friendship and the museum experience: The interrelationship of social ties and learning." Unpublished doctoral dissertation, University of California, Berkeley.

Samis, P. (2007), "New Technologies as a Part of a Comprehensive Interpretive Plan" in *The Digital Museum: A Think Guide,* Din, Herminia and Phyllis Hecht, eds., Washington, DC: The AAM Press, American Association of Museums, 2007.

Sayre, S. & Wetterlund, K. (2008) "The Social Life of Technology for Museum Visitors," Visual Art Research Journal, Pennsylvania State University.

Sayre, S. & Dowden, R. (2007), "The Whole World in Their Hands: The Promise and Peril of Visitor Provided Mobile Devices" in *The Digital Museum: A Think Guide,* Din, Herminia and Phyllis Hecht, eds., Washington, DC: The AAM Press, American Association of Museums, 2007.

Sayre, S. (2005), "Multimedia that Matters: Gallery-based Technology and the Museum Visitor," First Monday, Peer-Reviewed Journal on the Internet 10, #5 http://www.firstmonday.org/issues/issue10_5/sayre/index.html

Sayre, S. (1993). "The Evolution of Interactive Interpretive Media: A Report on Discovery and Progress at the Minneapolis Institute of Arts" in Diane Lees, ed., *Museums and Interactive Multimedia: Proceedings of the Sixth International Conference of the MDA and the Second International Conference on Hypermedia and Interactivity in Museums,* Museum Documentation Association and Archives and Museum Informatics.

Sachatello-Sawyer, B., Fellenz, R., Burton, H., Gittings-Carlson, L., Lewis-Mahony, J., & Woolbaugh, W. *Adult Museum Programs: Designing Meaningful Experiences.* Walnut Creek, CA: AltaMira Press, 2002.

## Glossary Terms

Prepared by Titus Bicknell, Ted Forbes and Nancy Proctor

### Alias/Shortcut

An alias or shortcut is an icon that enables one-click access to an application, program or document. It is possible, for example, to create an alias to a mobile website and store the link to it as an icon on the smartphone screen to provide more rapid access than typing in the website's URL. Apps are also generally represented on the smartphone screen by icons that are shortcuts to start up the app.

### Alternate Reality Game (ARG)

Alternate Reality Games are participatory, multi-platform experiences that may use a wide range of media and tools, including websites, mobile devices, and analog platforms in a physical space. Generally, the narrative of ARGs is shaped by the game players and their actions and movements through space and across platforms.

### Android

A mobile operating system based on Linux; although now owned by Google, Android is extensible by the vast developer community and many variants of Android have been deployed by different hardware manufacturers on mobile and portable devices such as smartphones and tablets.

### API

Common abbreviation for application programming interface, a set of defined parameters and interaction that allow different apps to communicate with each other. Twitter API allows other apps to collect tweets, query twitter accounts or send tweets; Facebook API allows other apps to post status messages, retrieve user account information or posted images.

## App

Common abbreviation for *application program,* which refers to any body of code that performs a task or tasks when installed on a given *operating system,* e.g., word processing, image manipulation or even a game. See *web app* and *native app* for specific uses pertaining to mobile devices in museums.

## Asset

An asset is a piece of content in a discrete, self-contained form such as an image, a video, an audio file, or a text document.

## Augmented Reality (AR)

Augmented reality is the "real world" overlaid with digital content to create a multi-sensory experience. Audio tours have been described as the original augmented reality, since the user's understanding or experience of a visual scene or environment can be "augmented" with audio heard at the same time. More commonly today, augmented reality is a location-based service delivered through a smartphone or tablet computer: the user views the object or scene in front of him or her through the camera on the device, and the screen shows that "real world" view overlaid with pertinent images or explanatory text. A common example of AR is found in directory apps that label the scene observed through the mobile device's camera with further information about restaurants, stores, etc., in the nearby area. In the cultural space, AR has been used to overlay contemporary environments with historic photos and site-specific digital artwork. AR depends on a location-based technology, e.g., GPS, to trigger the display of the correct content for the scene viewed through the user's device.

## Banner Ad

A banner ad is a horizontal advertisement that usually appears at the top or bottom of a web page. They are used in some mobile websites and apps to generate revenue and promote products and services.

## CMS

Common abbreviation for Content Management System, often deployed as a *web app* or *SaaS* to allow for collaboration among many users in creating, editing and organizing content that will be delivered via the Internet as a website, mobile website or delivered via a web app or published as part of a native app.

## CROWDSOURCING

A term popularized by *Wired* editor Jeff Howe in 2006, crowdsourcing refers to a collaboration with a broad user-base or general public to accomplish specific tasks. A museum might "crowdsource" photographs or geodata of objects and locations of interest, for example, by building a mobile app that allows users to capture this content and add it to a common data set from their personal mobile phones. Although the "crowd" might have specialized knowledge, skills or interests, participants are not individually recruited for a crowdsourced activity; rather they volunteer their services and may even be anonymous or unknown, contributing via a platform provided by the sponsor/developer of the crowdsourcing project.

## DAMS

Common abbreviation for Digital Asset Management System, often deployed as a web app or SaaS to allow for collaboration among many users in ingesting, tagging, and organizing digital assets. DAMS focus on secure storage of hi-resolution originals that can be leveraged by a CMS in platform-specific, reduced quality, alternative-format instances. This dynamic relationship between DAMS and CMS minimizes the number of digital versions of the same asset that need to be stored.

## DEVICE

Device is a term used to describe computer hardware. Devices include computers, phones, game systems, media players and other physical electronic aides.

The term "mobile" device usually refers to a handheld, portable piece of equipment, such as a smartphone, smaller media player or tablet.

## Drupal
An *Open Source* application widely used within the museum community as a *CMS*.

## eBook, eBook reader
An eBook reader is a portable device designed primarily to present text, such as books, magazines and article, i.e., make them readable in a digital format as "eBooks". E-Book readers often have some interactive and Internet capabilities, e.g., the ability to look up the definition of a word by tapping it on the screen, bookmark, annotate and highlight text. Some eBooks are simple PDF documents with text and images only; others may include video, audio, and more complex combinations of media.

## Geotag
A geotag is the x, y, and z coordinates or latitude, longitude and altitude of a given location. This metadata can be associated with buildings and outdoor points of interest through a process called "geotagging": collecting the geo-coordinates of the point and connecting them with relevant records or other content.

## GPS
GPS stands for Global Positioning System. It is a "line-of-sight" location-based technology that uses satellites to identify and relay the user's geo-coordinates (latitude, longitude, altitude) to his or her mobile device. GPS only works with any degree of accuracy outdoors, but because it requires a direct "line-of-sight' from the satellite(s) to the mobile device, it can function less well outdoors if obstructed by tall buildings or foliage

## HTML5

HTML is a common abbreviation for HyperText Markup Language which refers to the simple yet powerful *standard* that governs the writing and rendering of web pages; ratified by the W3C (World Wide Web Consortium), HTML5 is the latest iteration of the standard and adds support for the presentation and control of and interaction with audio, video and other visual assets without needing additional plug-ins. It is hoped HTML5 will allow the development of richer interactive content to run on mobile and portable devices with minimal processing power or ability to support plugins such as Flash.

## Hybrid app

A hybrid app combines features of a native app with web-delivered content and/or web pages.

## Interface

An interface is a visual, graphic representation (also called a GUI or graphic user interface) that provides clickable access to content and services on a digital screen.

## iOS

Apple's operating system for the iPhone and iPad; a powerful and feature-rich OS, but with limited extensibility by the developer community. Apple argues the tight controls it places on iOS ensure the highest quality of user experience; others argue that the restrictions hinder development and cite open-source platforms where large disparate communities collaborate to develop and optimize new functionality as a better model.

## Metadata

Refers both to data about data and data about the framework in which that data is contained. In most cases metadata is invisible to a user during their experience of data, but is used to inform or organize the user experience. Metadata can be used to group results in a search by key shared metadata not explicit in the search query, or metadata can be used to find associated assets in multiple CMS or DAMS in order to generate a rich media experience of a given piece of data.

## MOBILE GIVING

Mobile giving is a way of donating small amounts of money, usually $5 or $10, through text messaging (SMS) from a cellphone. The amount of the donation is added to the donor's mobile phone bill.

## MOBILE WEBSITE

A website optimized for access via a mobile device rather than a laptop or desktop computer. Mobile websites are formatted to be viewed on a range of small screens and include minimal graphics or media elements that require significant bandwidth to download or plug-ins to display correctly to the end user; some mobile websites are designed specifically to be experienced on certain devices, while others make use of different CSS sheets to delver a different layout, depending on what device the web server identifies as being used to visit the site.

## MOORE'S LAW

First described by Intel co-founder Gordon E. Moore in 1965, it indicates that the number of transistors that can be placed on an integrated circuit doubles every 24 months. The increase in processing power has in fact doubled every 18 months due to complementary developments such as parallel computing that support multi-cored processors and multi-threaded processing, to allow for multiple processors in the same computer.

## NATIVE APP

Refers to any application program designed to be installed and run on a specific operating system. It is most commonly used to distinguish between a user experience that runs natively on a device, usually without needing access to the

Internet, and a website that is formatted to feel like an app rather than a mobile website, and is accessed via an app icon on the user's mobile or portable device.

## NETWORK EFFECTS

"Network effects" are used to describe systems that become more useful or intelligent the more participants, elements or nodes are in the system. A common example is the cellphone, a prized possession precisely because it enables connections to such a large percentage of the people on the planet. A system may also achieve network effects by mixing elements or nodes in the network: e.g., an app may become more effective if it links to a mobile version of a museum's website for certain features, and is in turn promoted by the museum. In this example, the value of the whole system is greater than the sum of its individual parts.

## ONBOARD

"Onboard" refers to the local memory of a device: an "onboard" app or audio tour is stored in the local memory of a personal media player and can be played back without the device being connected to the Internet. Onboard content stands in contrast to "streamed" content.

## OPEN SOURCE

Open Source Software (OSS) describes any application where the source code is available freely to anyone for use and improvement. Many open source applications have attracted large dedicated communities whose cumulative improvement and extension allow for faster version release, and feature addition than many paid applications. While there is no purchase price for open source applications they are often wrongly described as free: considerable technical and staff costs may be required to make effective use of an open source application, but doing so often allows for cross-institutional collaboration and cheaper customization compared to paid apps. The museum community has made extremely good use of such Open Source applications as Drupal and WordPress.

## OS

Common abbreviation for Operating System, the foundation layer that is installed when a device is switched on, required by all installed hardware, applications and peripherals for their individual functionality, e.g., Microsoft Windows 7, Mac OS X, Linux on personal computers or Apple's iOS, Microsoft Windows Mobile, Android on mobile and portable devices.

## Personal Media Player

A personal media player (PMP) is a mobile, often handheld device that plays rich media such as audio and video from local memory on the device. A common example is the iPod. PMPs may or may not have Internet access (wireless) capabilities.

## Platform

Platform can refer to a) the operating system running on a specific piece of hardware (Windows 7 is a platform), or b) a method of distributing content (Facebook, YouTube and an iPhone app are all platforms we can use to distribute content).

## QR code

A QR code is a kind of bar code that, because of its more complex design, can store more information than a traditional bar code. QR codes are read by dedicated apps that use the smartphone's camera to decode the content "stored" in the code's pattern. This content is often a URL or website address, so upon reading the QR code, the code reader app will attempt to display the web page at the given URL. Because of their higher content storage capacity, QR codes can also transmit up to several thousand alphanumeric characters, e.g., contact information or short texts.

## Smartphone

A "smartphone" is a phone that has Internet connectivity, enabling it to provide access to websites, apps and downloadable content. Smartphones are distinguished by simpler, often earlier, mobile phones that provide only voice calling and text messaging (SMS).

## SMS
Short message service or "text messaging" allows messages of up to 160 characters in length to be sent or received from a cellphone.

## SOCIAL MEDIA
Social media are platforms built on the Web 2.0 model that enable and encourage the co-creation and exchange of content, conversations and shared knowledge and experiences. Common examples include Facebook, Flickr, Twitter, YouTube and blogs.

## SPECIFICATION
Sometimes a synonym for *Standard* but referring to a principles, coding vocabulary or mark-up language that is under consideration as a standard; multiple specifications originating in different development contexts can often be aggregated to define a *Standard* that multiple developers can then adhere to and leverage.

## STANDARD
Refers to a set of ratified principles, coding vocabulary or mark-up language used to organize and manage content, its appearance and data structure. The adoption of standards ensures wide compatibility with devices, operating systems and applications developed by different manufacturers.

## STREAMING
"Streaming" refers to the download and playback of media in real time, without storing a copy of the content in the memory of the device. Unlike "onboard"

content delivery, streamed content requires a constant Internet connection in order to play media. Quite often streaming is used when the content owner does not want users to be able to keep a copy of the content.

## Swiping

"Swiping" is a touchscreen gesture introduced by Apple with the iPhone. A lateral swipe or drag of the finger across the screen either left or right will cause a menu of content, often images, to flow by at a speed that responds to the speed of the gesture.

## Tablet Computer

A tablet computer or simply "tablet" is a portable digital device that consists of a screen with all computing components built behind the screen. Tablet computers do not have keyboards built-in. A common example of a tablet computer is the iPad.

## 3D

3D stands for three-dimensions, and usually refers to digital models and assets that represent objects and environments in full three-dimensional form or space. compass (in a smartphone)
Many smartphones now include digital compasses that will indicate the bearing of the device as held by the user. This functionality can be used by apps run on the smartphone to perform and inform more complex spatial operations, e.g., help navigate the user through a space by recognizing not just where the user is, but also the direction in which s/he is heading.

## Web App

Common abbreviation for *web application program* and also know as *SaaS (Software as a Service)*, which refers to any body of code that performs a task or tasks when installed on a given Internet-accessible server and a user interacts with it via a web browser. The key differences from an *app* are that it does not require installation on the user's computer and the processing power required to execute the task(s) can be leveraged from a number of servers in the cloud. A key advantage is that the web app can be upgraded for all users by updating the central server app rather than each individual user's locally installed copy; the key disadvantage is that the user requires an internet connection to use a web app. N.B.: in discussions of iOS and Android devices, web app has come to refer also to mobile websites that are accessed through an app icon on the user's device; they feel like a native app but rely on the content and experience being delivered via the Internet.

## WEB 1.0, WEB 2.0

Web 1.0 is a shorthand to reference a style of content and experience design that was common in the early days of the Internet: also known as a "one-way" or broadcast model, Web 1.0 delivers messages to end users but does not "listen" or receive feedback from users. By contrast, Web 2.0 signifies a two-way communication model, with an exchange of content and messages between the author/ broadcaster and audiences. Web 2.0 is generally heralded as a less formal, more conversational approach to digital (and other) content and experience design, soliciting audience response and partnership in the dialogue. Social media is "Web 2.0" in nature and concept.

## WIRELESS

Wireless generally refers to a network connection that does not require any 'wires' or cables other than, perhaps, to a power supply. Wi-fi and 3G are common wireless network access protocols.

## XML

Common abbreviation for Extensible Markup Language, a W3C-ratified standard for organizing data. Due to its flexibility and ability to support custom-defined containers, it has become a de facto standard for passing data among different incompatible applications.

# Authors

TITUS BICKNELL is a technologist, co-founder of pink ink http://www. titusbicknell.com/pinkink and TheGalleryChannel.com, and co-principal of museummobile.info. Apart from a fascinating stint at NBC Universal in 2007–08 working on the big screen, Titus has spent the last 10 years exploring the small screen, both web and hand-held. As Chief Engineer at Antenna Audio and subsequently Head of Mobile Technologies at Discovery Communications, he was fortunate to participate in ground-breaking hand-held projects at Tate Modern, the Louvre, The Centre Georges Pompidou, the Intel Museum, and the Getty, among others. As Director of Business & Production systems at Hendricks Investment Holdings, he developed production workflows for documentary and publicity programming, live sports event in-stadium broadcasting and asset optimization for mobile and portable app development, as well as overseeing the Property Management, IS and online systems for Gateway Canyons Resort, Gateway Colorado Automobile Museum, and Discovery Retreats. He is currently Chief Engineer at MyDiscovery, a new division of Discovery Communications. At various times Titus has been a filmmaker, Latin scholar, avid cyclist and fund raiser for the Lance Armstrong Foundation, and a plug-in developer for the WordPress open-source platform

ALLEGRA BURNETTE is the Creative Director of Digital Media at The Museum of Modern Art, New York, overseeing the design and production for the museum's website, MoMA.org, as well as mobile devices, interpretive kiosks and digital displays. Online projects include two complete site redesigns, creating the online collection and audience-specific sites for teachers, teens, and kids, overseeing an ongoing series of award-winning exhibition sites, and extending the reach of MoMA's content through iTunes U, YouTube, mobile apps, and elsewhere. Prior to working at MoMA, Ms. Burnette created and ran a media department at the renowned museum exhibition design firm Ralph Appelbaum Associates. She has an MFA in museum exhibition planning and design from the University of the Arts, where she has also taught graduate courses in museum media. She currently teaches in the online graduate program for Museum Studies at Johns Hopkins University and serves on the board of the Museum Computer Network.

JANE BURTON is Head of Content and Creative Director, Tate, London. She leads the media team responsible for Tate's video and film productions, including documentaries about artists for TV and online, the weekly video podcast TateShots, and the recently announced Tate Movie, an animated feature film made by and for children, produced in collaboration with Aardman. In 2002, she launched the world's first wireless multimedia tours at Tate Modern, winning a BAFTA (British Academy of Film and Television Arts) award for innovation, and in 2008 piloted the UK's first gallery tour for iPhones at Tate Liverpool. She initially joined Tate in 1999 as Curator of Interpretation. She previously worked as a journalist and editor for UK national newspapers.

TED FORBES is the Multimedia Producer at the Dallas Museum of Art. His work spans the production of interactive and digital content including exhibition Web sites, teaching materials, in-gallery interactive content (kiosks and touch screens), and video production. He is currently working on two major projects for the Museum—DallasMuseumOfArt.tv, the online hub for museum multimedia content, and DallasMuseumOfArt.mobi, the in-gallery content distribution interface. Forbes has been an adjunct faculty member at Brookhaven College since 2003, teaching interactive and Web design. He began his work in the tech industry as a producer for iSong.com creating and producing music education software, and later led his own design studio, producing print and interactive content for many clients including Microsoft, Best Buy and Public Broadcasting, the Dallas Opera, the Science Place, the Illustrators Partnership of America, and RasGas LNG in Doha, Qatar.

KATE HALEY GOLDMAN recently joined the senior staff of the Center for Interactive Learning, Boulder, Colo., a nonprofit founded by the Space Science Institute. Prior to taking the position of Director of Learning Research and Evaluation, she was a Senior Research Associate at the Institute for Learning Innovation since 2000. Her work concentrates on furthering theory and practice of the use of technology in museums and related informal learning environments. She has directed projects both in the U.S. and abroad, involving exhibits and program evaluation, mobile phones and smartphone apps, websites, gaming, augmented and mixed reality, novel data visualization systems, and online learning. Recent projects she has directed include: "Making Space Social," a Space Science Institute Facebook game on origins, Space Science audience research for the Encyclopedia of Life, summative evaluation of the NSF-funded computer game WolfQuest, program-level nation-wide evaluation of NOAA's "Science on a Sphere," and the NSF mixed and augmented reality exhibition projects "Virtual Human" (Boston Museum of Science and University of Southern California Institute for Creative Technologies), ARIEL (Franklin Institute) and "Water's Journey" (University of Central Florida Mixed Reality Lab and Museum of Discovery and Science). Currently, she is Co-PI of the NSF-funded open source project, Open Exhibits. (OpenExhibits.org)

ANN ISAACSON is Associate Educator at the Minneapolis Institute of Arts. She has worked with children's arts programs and museum education the past 14 years as manager of Free Arts Minnesota from 1996–1998 and as coordinator of the Art Adventure Guide Program at the MIA, 1998 to present. The focus of her work is developing curriculum and training museum guides. Ann has a degree in Studio Arts and an M.A. in Design History from the University of Minnesota.

SHEILA MCGUIRE is Director of Museum Guide Programs at the Minneapolis Institute of Arts (MIA). She, department staff, and volunteers in three tour guide programs create interactive tour programs for diverse audiences that support the department mission to provide volunteer-facilitated learning experiences that inspire visitors to discover personal meanings in art and explore museums confidently on their own. She teaches and evaluates volunteers, oversees a Visual Thinking Strategies partnership with the Minneapolis Public Schools, and represents the museum's Division of Learning and Innovation on several cross-functional exhibition, programming, and interpretive teams. She received her M.A. in art history from the University of Minnesota and her B.A. in art history from the State University of New York at Purchase. Over the past 30 years she has taught adults and young people on a wide variety of subjects at the MIA, Lawrence University (Björklunden), Walker Art Center, University of Minnesota, and the Metropolitan Museum of Art in New York.

NANCY PROCTOR heads up mobile strategy and initiatives for the Smithsonian Institution. With a Ph.D. in American art history and a background in filmmaking, curation and art criticism, Proctor published her first online exhibition in 1995. She co-founded TheGalleryChannel.com in 1998 with Titus Bicknell to present virtual tours of innovative exhibitions alongside comprehensive global museum and gallery listings. TheGalleryChannel was later acquired by Antenna Audio, where Nancy headed up New Product Development from 2000–2008, introducing the company's multimedia, sign language, downloadable, podcast and cell phone tours. She also led Antenna's sales in France from 2006–2007, and worked with the Travel Channel's product development team. From 2008–2010 she was Head of New Media at the Smithsonian American Art Museum. Nancy is program chair for the Museums Computer Network (MCN) conference and co-organizes the Tate Handheld conference, among other gatherings for cultural professionals. She also manages MuseumMobile.info, its wiki and podcast series, and is Digital Editor of Curator: The Museum Journal.

ED RODLEY is the senior Exhibit Developer at the Museum of Science, Boston, with over 20 years experience in all aspects of exhibition development. He has developed major international exhibitions on the Soviet space program, Leonardo da Vinci, Egyptian archaeology, and using Star Wars as a way to look at new technologies. He has developed exhibitions, websites, audio tours, multimedia tours, augmented reality exhibits, and edited the National Geographic book Star Wars: Where Science Meets Imagination. Incorporating emerging technologies into museum practice has been a theme throughout his career. His work is deeply influenced by constructivist learning theory, but he's no zealot. He is active in the National Association for Museum Exhibitions (NAME), presents regularly at conferences, and blogs at Thinking About Exhibits. He also consults on interesting projects large and small for clients like Harvard University, The History Channel, Random House Books, the U.S. National Park Service, and others.

PETER SAMIS is Associate Curator of Interpretative Media at the San Francisco Museum of Modern Art (SFMOMA). In the early 1990s, he served as art historian/content expert for the first CD-ROM on modern art, and then spearheaded development of multimedia programs for SFMOMA's new building. Since then, SFMOMA's Interactive Educational Technologies (IET) programs have received awards from sources as diverse as the American Association of Museums, the National Educational Media Network, and I.D. Magazine. The Museum's 2001 exhibition Points of Departure was the first to introduce mobile multimedia devices into the galleries; it won AAM's Gold Muse Award. In 2006, SFMOMA won three more Muse Awards, including one for the podcast series "SFMOMA Artcasts," which also won "Best of the Web" in Museums & the Web's Innovative and Experimental category. Since that time, SFMOMA's IET team has partnered with NOUSguide and Earprint to develop and deploy Making Sense of Modern Art Mobile, a multimedia guide to the permanent collection, offered free to museum visitors, typically delivered on the iPod Touch.

SCOTT SAYRE is a founder and principal at Sandbox Studios, a Minneapolis-based group that works with museums to plan, create, manage and assess education programs and technology projects. He is also principal of Museum411, a sister company developing mobile audio solutions for cultural institutions. Scott has a doctorate in education and has over 25 years of experience working with emerging education and information technologies. He teaches at Johns Hopkins University, Università della Svizzera italiana, Lugano, and has previously taught media and technology design and planning at the University of Minnesota, Minneapolis College of Art and Design and the University of Victoria. In 2002-2003 Scott served as the Art Museum Image Consortium's Director of Member Services and U.S. Operations. From 1991–2002 he was the Director of Media and Technology at the Minneapolis Institute of Arts. Scott was chair of the American Association of Museum's Media and Technology Committee and is currently an Editor for Museum-Ed.org and serves on the New Media Consortium's board of directors. Prior to his work with museums, Scott was a Media Applications Researcher at the University of Minnesota's Telecommunications Development Center.

MARGRIET SCHAVEMAKER is an art historian, philosopher and media specialist. After a career as lecturer and assistant professor at the art history and media studies departments at the University of Amsterdam, she currently holds the position of head of collections and research at the Stedelijk Museum, Amsterdam. Schavemaker has written extensively on contemporary art and theory, co-edited several volumes (including *Now is the Time: Art and Theory in the 21st Century* (2009) and *Vincent Everywhere: Van Gogh's (Inter) National Identities* (2010)), and is an acclaimed curator of discursive events and public programs. In recent years, new media have been high on Schavemaker's agenda, resulting in a.o. the ARtours project and the creation of an augmented reality platform for smartphones, which can be used by museums to present their collection in innovative and interactive ways, inside and outside the museum.

KOVEN J. SMITH is the Director of Technology at the Denver Art Museum. With over a decade of museum practice, including stints at the Indianapolis Museum of Art and the Metropolitan Museum of Art, his primary focus has been the presentation of museum content in the digital domain. Koven is currently a member of the Museum Computer Network board of directors, founder and chair of the MCN Semantic WebSIG, and a former steering committee member of the steve.museum and ConservationSpace projects. He has also been a featured speaker at Ignite Smithsonian, the Museums and the Web conference, the American Association of Museums conference, and the Tate Handheld Conference, among others. He writes about museums and technology at http://kovenjsmith.com and Twitters at http://twitter.com/5easypieces.

ROBERT STEIN is the Deputy Director for Research, Technology and Engagement at the Indianapolis Museum of Art (IMA). In that role, Stein leads a wide range of activities for the museum, including an extensive effort in media, web technology and software. Since 2006, he has played a significant role in shaping the way that the IMA has applied technology and media to the mission of the museum. In 2009, Stein and colleagues launched the streaming video website, ArtBabble.org. Awarded the 2009 American Association of Museums MUSE Award for Best Online Presence and the 2010 Best Overall Site award from the International Conference of Museums and the Web, ArtBabble brings together 28 prominent cultural organizations to create a true destination for art video online. Also in 2009, the IMA created and released TAP, a mobile tour platform to help museums author and distribute mobile content, both in the galleries and online. Stein and colleagues have continued to advocate for open content standards and specifications within the museum community to support sustainable content. Stein currently serves as Secretary on the Board of the Museum Computer Network and is active in speaking and writing on topics related to museum transparency, technology and scholarly research.

KRIS WETTERLUND is a founder of the Museum-Ed Discussion List and the Museum-Ed.org website, created to meet the needs of museum educators by providing tools and resources by and for the museum education community. The Discussion List, begun in 1995 and the Web site, created in 2003, function as a not-for-profit organization that serves 1600 members of the Discussion List and the greater community by providing resources on the Museum-Ed.org Web site that are free and available to educators in any type of museum, and anyone interested in the field of museum education. Kris is currently an editor for Museum-Ed.org. She is also a founder and principal at Sandbox Studios, consulting with museums to plan, create, manage and assess education programs and technology projects. She has extensive experience working with teachers and art museums and has designed and implemented a two-year program to train K–12 teachers throughout Minnesota to use online art museum resources and technology in their classrooms. She has worked as an art museum educator for the past 18 years, in the education department at the Minneapolis Institute of Arts from 1990–2000 and as Director of Education at the Minnesota Museum of American Art from 2000–2005.ucation at the Minnesota Museum of American Art from 2000–2005.

# Index

# M

# INVEST in YOURSELF

"I want to be the very best in my field and these resources will provide me with the information I need to kick-start my professional development."

Patricia Counihan, Concord Museum, MA

AAM membership gives you access to information for success in your job and career.

- Year-round professional development opportunities
- Timely updates on what's happening in the field
- Online library with thousands of indispensable resources
- Connections to professional networks
- Resources on standards and best practices
- Discounts in the AAM Bookstore
- Alerts to job opportunities and fellowships

## Join today!
www.aam-us.org/joinus or 866-226-2150

AMERICAN ASSOCIATION OF MUSEUMS